RIDING, ROPING, AND ROSES

Colorado's Women Ranchers

By Judy Buffington Sammons

WESTERN REFLECTIONS PUBLISHING COMPANY®
Montrose, CO

© 2006 Judy Buffington Sammons
All rights reserved in whole or in part.

ISBN-13: 978-1-932738-29-2
ISBN-10: 1-932738-29-0

Library of Congress Control Number: 2006923895

First Edition
Printed in the United States of America

Cover and text design by Laurie Goralka Design

Western Reflections Publishing Company®
219 Main Street
Montrose, CO 81401
www.westernreflectionspub.com

for my son
Scott
who has what it takes

Acknowledgments

I wish to first of all thank P. David and Jan Smith, owners of Western Reflections Publishing Co., for putting *Riding, Roping, and Roses* into print. They, along with their Managing Editor, Carole London, and my editor, Bonnie Beach, did a splendid job of melding my stack of papers and pictures into a real book. I also acknowledge gratefully the help of many other people who assisted me. These are credited in the interview sections of the chapter bibliographies at the end of the book. Credit is also due to many librarians around the state, who willingly and pleasantly searched out obscure newspaper articles and pictures for me.

    Special thanks must also go to several people for their substantial help with individual chapters. The chapter on Marie Scott would not have been possible without the help of Marie's fine friend—and now mine—Mario Zadra. Mary Wood of Ouray also helped greatly with the chapter on Marie. Donna Wilson, Executive Officer of Cherokee Ranch and Castle Foundation, was instrumental in helping me accurately portray Tweet Kimball. Thanks to the Delta County Historical Society and photographer Ben Walker for sharing their information on Genevieve Hardig. Jerry and Polly Bates introduced me to Creede and Grant Houston, Editor of the *Lake City Silver World*, provided the pictures that helped bring Mabel Steele Wright into the light. Janice Nixon, an old highschool friend, and two of Mavis Peavy's granddaughters, Jennie Mai Bonham and Marsha Witte, are responsible for much assistance with the chapter on Mavis. Eileen Burt Hanson, descendant of the Becker and Hanson cousins, shared her home and her extensive family records and photographs with me, and the Monte Vista Historical Society provided the famed O.T. Davis photographs that added so much to the chapter on those long ago cowgirls of the San Luis Valley. Thanks to the Museum of Northwest Colorado for sharing their photographs of the Bassett family. Suzanne Esty, granddaughter-in-law of Vevarelle Esty, is responsible for information and pictures in that chapter. A big thank you to photographer and friend, Walt Barron, for his excellent photograph of Barbara East. Special thanks to my son, Scott Stoneburner, for a great day spent searching for Barbara East on the top of Grand Mesa.

Table of Contents

Introduction 1

Chapter One   HAVING WHAT IT TAKES 5

Chapter Two   MARIE *Marie Scott of Ridgway* 17

Chapter Three   TWEET *Tweet Kimball of Castle Rock* 35

Chapter Four   GENEVIEVE *Genevieve Hardig of Delta* 47

Chapter Five   MABEL *Mabel Steele Wright of Creede* 57

Chapter Six   MAVIS *Mavis Peavy of Padroni—Logan County* 71

Chapter Seven   EMMA, ANNA, AND LIZZY—LIZZIE AND EMMA . . 83
The Becker and Hansen cousins of San Luis Valley

Chapter Eight   ELIZABETH *Elizabeth Bassett of Brown's Park* 95

Chapter Nine   BARBARA *Barbara East of Boulder* 107

Chapter Ten   VEVARELLE *Vevarelle Esty of Gunnison* 117

Bibliography 129

Introduction

When I was a young girl growing up in the fifties on a ranch in Gunnison, Colorado, the legendary Marie Scott—a small, red-headed, bow-legged ranch woman with a Stetson hat sitting well back on her head—arrived one day with her neighbor, Mario Zadra, to buy bulls. My grandfather, Clyde Buffington, was a Hereford breeder and often had out-of-town buyers stopping at the ranch to look over the range bulls, but these buyers were always men. A woman rancher of Marie's status was practically unheard of at that time. She was a legend of sorts for having gained a foothold in a very male dominated business—an almost impossible feat. Not only had she become very well-known as a rancher, but she was well on her way to becoming one of western Colorado's largest land owners. And so, when Marie Scott came to call at our ranch, we all took notice. After discussing the bulls' genetics with my grandfather and making her selection, Marie built a fire in our ranch yard, branded her bulls, paid for them, and left. She had certainly captured my attention.

Later in life, I became acquainted with another equally remarkable woman rancher. I discovered, on the old ownership records of ranch property that my family owns, that our place had been homesteaded by a woman. Women homesteaders were rare in the settlement of the Gunnison Country, but certainly not unheard of. The old-time ranchers in my valley had always referred to our property as the "Missionary Place"—named for the missionary who had homesteaded it, whom everyone assumed was a man. On finding that the missionary was a woman, a feeling of kinship developed and, as the present owner, I felt compelled to find out as much as I could about the mysterious woman who went before me.

I discovered that our missionary had a lovely name—Rosalia. Her homestead was beautiful—located in the northern part of Gunnison County in the Ohio Creek Valley, on the fringes of what is now the West Elk Wilderness. Rosalia Keever Grahm could not have picked a more scenic place—or a more difficult one, as far as homesteading goes. One could easily second-guess her choice of land and say that she had chosen unwisely and was doomed to fail. Her 160 acre homestead was located at an elevation of nearly 10,000 feet, where snow frequently piled up to four- or five-foot depths in the winter, making the place nearly inaccessible for several months of the year. Stores and towns were remote and roads rutted, rough, and muddy in the summer months, making wagon travel difficult at best. Rosalia somehow survived on the homestead, in part by keeping milk cows and selling cream, milk, and butter to the nearby coal mining town of Baldwin. She tried to keep her missionary work alive, too, and was frequently seen on the streets of Gunnison, playing a musical instrument and collecting money for her religious cause—whatever that may have been. How she and her teenage sons coped with the solitude of the Missionary Place, the fear of failure, and countless other problems beyond their control is something one can hardly imagine. When they couldn't get out of the place in the deep of winter—which is a likely possibility—the aching sense of isolation, the confinement, and bitter cold temperatures must have been a trial to test even the toughest pioneer's resolve.

Rosalia's years on the homestead were remarkable, but few—she stayed just long enough to fulfill the requirements of the Homestead Act, then she sold the property and departed from the Gunnison Country, leaving no record as to the reason. The remains of her log cabin and the rhubarb she planted nearby are all that are left of Rosalia's homesteading venture. I never walk by the rotting logs of her cabin and the rusty pieces of her cookstove that now lie on the ground without thinking of her and her gallant effort to become a rancher.

My earlier encounter with Marie Scott, as well as the intriguing information I had discovered about Rosalia, stayed with me for years, and, later in life, when I began a writing career, it was in the back of my mind to write about them. Doing research on other history projects, I discovered that Rosalia and Marie weren't the only bonafide women ranchers in the state. There were others like them, but very few. In fact, there were only a handful of women who had earned reputations as genuine and successful women ranchers, and they fascinated me. I set about trying to learn more about them and the circumstances that had shaped their lives, and then try to put their stories into words. In doing so, I had to deal with the fact that Marie Scott, perhaps the most well-known of them, had little use for writers and had no wish to be written about. Indeed she once discussed writers with her friend and business associate, Mario Zadra, stating: " . . . anyone with that kind of time on their hands should get those hands on a shovel."

That aside, writing about Marie and other Colorado women ranchers—Castle Rock's Tweet Kimball, Delta's Genevieve Hardig, Creede's Mabel Steele Wright, Logan County's Mavis Peavy, the San Luis Valley's Hansen and Becker cousins, Brown's Park's Elizabeth Bassett, Boulder's Barbara East, and Gunnison's Vevarelle Esty—had another particular kind of difficultly. All these women had been so unique in their own communities that they had become legends, as Marie Scott had, and the stories about them had grown with time and often had become embellished with the telling. Sometimes it was hard to separate the myths from the truths. My own background, as a rancher's daughter and granddaughter and now owner of some of the family ranch property, helped me to sometimes recognize a tale too tall to be true.

Researching the ranching history of my own hometown of Gunnison, Colorado, for my first book, *Tall Grass and Good Cattle—A Century of Ranching in the Gunnison Country*, got me thoroughly hooked on visiting the old glory days of ranching. Doing the research for *Riding, Roping, and Roses* enabled me to continue with this exploration and expand on it. I greatly enjoyed my journey around the state to become acquainted with other ranching communities.

Sometimes considered an "outsider" at first, it was never long before I came to know a new area's geography and history and, best of all, its people. In every town, ranch people generously shared their time and their country with me. We relived ranching history in old pickup trucks, on horseback, over kitchen tables, in restaurants with cowboy motifs, high in the mountains in remote cow camps, by letter, email, and telephone. I gathered vital and interesting information all along the way and made many valuable and lasting friendships, as well.

My admiration for the women ranchers I researched only grew as I visited their country and learned more about them. What a privilege it was to be in their company, as was the case with Vevarelle Esty when we roamed around the Esty ranch shortly before her death. I began to feel that I was getting to know Marie Scott just a little bit while digging through the letters and pictures provided by her good friend and heir, Mario Zadra, and while sitting around a ranch house kitchen table with one of her contemporaries, Willma Potter of Ouray, Colorado. Walking what was once Genevieve Hardig's sheep ranch and inspecting photographs of her, now housed at the Delta County Historical Society, was like getting to know this remarkable lady. I shared lunch in the rain with Barbara East high up on Grand Mesa—an interview conducted in a truck. Visiting Mabel Steele Wright's grand old ranch house in Creede was like stepping back to the time when she was there herself, as her home has been little changed. And with each of the women I wrote about, I became acquainted with their close friends and neighbors—people of strong and genuine ranching heritages. By connecting with their friends and, in some cases, the women themselves, I became as intimately acquainted with these women ranchers as was possible.

I felt a close kinship with the people I wrote about—the remarkable women of *Riding, Roping, and Roses*. I chose that title to represent the rare combination of abilities that these women possessed: manly cowboy skills with a touch of femininity. They all made their concessions—from Marie Scott's "war paint" rouge to Vevarelle Esty's lush garden and home-canned vegetables to Tweet Kimball's mink coat. It was a pleasure to examine their lives—these great ladies, some of whom were born into the world of ranching on the last fringes of the pioneer period, some at mid-century, and some in the present day, and all of whom survived on steady diets of very hard work. I liked the enthusiasm and the fire with which they lived. In their later years, their ranching days behind them, they still maintained a fierce passion for the lives they had lived. In time their contemporaries—the cattlemen and sheepmen—came to respect them—maybe differing with their opinions or their methods, but in the end, not with their authenticity.

In his book, *Charles Goodnight—Cowman and Plainsman,* Ebetts Haley credits Charles Goodnight, the old-time Texas trail driver, with saying about cowboys, "Timid men were not among them—the life did not fit them. I wish I could convey in language the feeling of companionship we had for one another. Despite all that has been said of him, the old-time cowboy is the most misunderstood man on earth. May the flowers prosper on his grave and ever bloom, for I can only salute him—in silence." Charles Goodnight, one of my heroes, said it well of the cattlemen. I'd like to travel back in time to Texas and sit with Mr. Goodnight at his kitchen table and tell him what was going to happen next—what was going to happen with the women. I can almost see him smile.

Chapter One

HAVING WHAT IT TAKES

> *"Persons afraid of coyotes and work and loneliness had better let ranching alone."*
>
> Wyoming rancher, Elinor Stewart, *Letters of a Woman Homesteader*

In the old West, around the turn of the century, a few ranchers' daughters—a brazen few—decided to shake up the establishment a little bit, "rock the boat" and "rattle a few cages." They put on shocking divided skirts they had stitched up themselves or pants they had borrowed from fathers or brothers. They abandoned their ridiculous sidesaddles and dared to get on their horses astride. Then they happily rode off, leaving their ladylike images in the dust. They shot coyotes in Montana, rode the range in Wyoming, homesteaded in Nebraska, roped steers in Nevada, and branded mavericks in Colorado. A brave few of them—with a new taste of freedom— kept at it, weathering their faces, hardening their bodies and maybe their minds, shocking their neighbors, and along the way developing the same passion for the cowboy way of life that men had. They married or didn't—they inherited or homesteaded or bought ranches. In the very Western and very male world of cattle ranching, they became bonafide ranchers—America's first women ranchers: a new breed.

Colorado's early-day women ranchers—at least those who stayed with it and made a success of it on their own—were few and far between. Most were daughters of pioneer ranchers who had homesteaded around the turn of the century either in Colorado or a nearby state. Many had a strong female role model somewhere along the way to serve as their example. They were young widows of ranchers or divorced from ranchers or were never married and, therefore, pretty much on their own. Many of them had hands-on experience with day-to-day ranching chores from an early age and stayed with ranching all their lives, in spite of great difficulty. They not only

Chapter One Having What It Takes

The rancher's wife on the left wears the customary cowgirls' attire of the 1920s.

Photo courtesy of Bill Sanderson.

managed their ranches but, in many cases, actually did much of the physical work themselves.

Among the ranks of Colorado's first women ranchers was Margaret Duncan Brown, who came to be known as "The Shepherdess of Elk River Valley." Her memoirs were published in a book that bears that same title.

Her lifetime of ranching took place near Steamboat Springs. For forty-seven years, she lived alone on a sheep ranch that she ran on her own after her husband died from the flu epidemic of 1918. Margaret had no experience to bring to sheep ranching, having been raised a Southern belle before coming west shortly after her marriage. In her wedding picture, taken in 1900 at the age of eighteen, Margaret is an attractive and fashionable young lady in a Gibson girl gown and a large hat—under it soft brown curls, intelligent eyes, and a sweet smile. In 1915, a naive Margaret and her equally greenhorn husband decided to give ranching a try and bought 160 acres in the beautiful Elk River Valley, north of Steamboat Springs. After her husband's unexpected death, Margaret made a courageous decision to continue on with the

Annie Sammons, 1920s cowgirl of Powderhorn, Colorado, dressed in angora chaps.

Author's collection.

Chapter One Having What It Takes

Rosalia Grahm, early homesteader in Gunnison, Colorado, "proved up" on 160 acres located just beyond this hay field. Walt Barron photo. Author's collection.

Another example of women ranching in the Gunnison area is that of the eight Vidal sisters—daughters of pioneers who managed the family ranch after the deaths of their parents. Here, seven of the sisters are pictured in their "overalls" in the late 1800s.
Photo courtesy of *Gunnison Country Times.*

venture they had started together, even though she expressed great misgivings about her ability to carry on alone. She claimed in her memoirs, "My original ignorance of the real facts of farm and ranch life were appalling . . . And so I struggled on . . . I did not want to go back to town . . . difficulties continued to mount . . . I would need courage, which is that special kind of fear." After her timorous start, Margaret Duncan Brown persevered, devoting the rest of her life to the operation of the sheep ranch. She payed off the original 160 acres and enlarged it to 713 acres by the time of her death.

Women such as Margaret Brown were the exception in the early part of the 1900s. Most women of this time period had little choice in occupation other than the traditional one of wife, mother, cook, housekeeper, and helpmate. A career before marriage was not unusual, but it was almost always brief and generally that of a schoolteacher or nurse. Marriage cut careers short, and whatever other talents women may have possessed they were often not realized. Most of the early-day ranch wives fell into the traditional "helpmate" category. Women's work (and men's, too, for that matter) was staggering during these years. Intense physical labor was demanded from sunup to sundown, and women often performed theirs with the added burden of caring for many children and with constant pregnancies. They made their own soap and candles, sewed clothes by hand or on crude machines, kept the old "Majestic Range" going with the constant feeding of coal or wood, and planted and weeded huge gardens, then canned the produce for the coming winter. Pioneer women often summed up their early experiences on ranches in one word—"work."

Life got a little easier for ranch wives by mid-century, and although their role as ranch wife was still important, it remained a traditional one. For example, a ranch wife's typical day in August during haying season would begin very early with the cooking of breakfast for a large crew. After breakfast and breakfast dishes, she would clean the house, do the laundry, mend clothes, run to town for groceries and maybe tractor parts, then return home in time to cook the haying crew's evening meal. There were no "convenience" foods, microwave ovens, or clothes dryers. Often a daughter was the main dishwasher—put to work against her will—all the while longing to be out in the hay field with her brothers, where life was more exciting. Once in awhile, a wife took a more active role in the physical labor of the ranch—working in the hay fields or helping drive cattle on occasion—but

Chapter One Having What It Takes 11

this was rare. It was understood by all, including young girls, that the main role of ranch women was to be a domestic one.

Although this domestic role had its rewards, and many rancher's wives enjoyed life on a ranch and remember it with fondness, there were a few "cowgirls" around who simply had no taste for it. They wanted their days filled with different kinds of tasks than those of the typical ranch woman. On the same summer day earlier described for a rancher's wife, a woman ranching on her own might be doing things quite differently. She might start by milking cows and changing the irrigation water in the south "forty," followed by loading up salt blocks in an old Jeep to take to the cattle in the hills, then come back home to saddle up "Old Becky" and herd cattle into a higher pasture to new grass. For those who had chosen to be women ranchers, this kind of life seemed to have more freedom to it. These women experienced the wide-open spaces in doing ranch work—they trailed cattle, branded calves, and camped out on the range at roundup time. One of the women in this book, Vevarelle Esty, started training the ranch's horses as a child. By the time she was a teen,

A very young Vevarelle Esty in a divided skirt, trick riding at her father's ranch.
Photo courtesy of the Esty family.

she was somewhat practiced at being a trick rider. She and the others took to these kinds of activities as naturally as other young women of their time took to learning the skills of housekeeping.

The early-day cowgirls discovered they could be as good at many of the cowboying skills as a man. At age sixteen, Ridgway's Marie Scott ordered a pistol—a .32 revolver—out of a catalogue. She practiced with it until she became a crack shot—made sure this was known around—and then slept with it under her pillow. It was a new and powerful feeling to be regarded as a crack shot, and—undoubtedly—there were other such feelings to be discovered, too. Marie and the other newly coined cowgirls had rich and vital kinds of adventures that more traditional women rarely had a chance to experience. They rode horses through the sagebrush after the rain and became intoxicated with the smell of it—the fragrance of the West. They camped out under the stars, and saw the fiery sunrises of daybreak, and warmed their hands on tin coffee cups by campfires. They saw grass turn into beef and beef turn into money, and they knew that even more grass to feed more cows could turn into more money, and that money could turn into independence. And although they knew there would be good and bad years in the livestock business, they started believing they could tough them out. Then they were hooked, and there was no going back—no going back to the kitchen.

While the early-day cowgirls may have loved their new-found freedom, it was never going to be easy for them. Ranching was a demanding and difficult job, and sometimes it could be dangerous. A trusted horse could startle as a sage grouse flew suddenly out of the brush, leaving its rider unconscious and alone far from home. Along with the hardships of sheep or cattle ranching, the women still faced the prejudice and bias that was especially strong in the first half of the century against females in nontraditional roles. They may not have been invited to join the local cattlemen's organization, where "good old boys" did favors for one another. Some of them got their start in ranching before women even had the right to vote. And they soon found that lacking a man's physical strength was a problem when ranching required that kind of brawn. Even some men had difficulty standing up to the rigors of ranching. In time, most ranchers acquired a very muscular, athletic, and wiry look, accompanied with a leathery face and a bow-legged gait. Well into their seventies and eighties, an old-time cattleman would retain the hard muscled, weathered, and eagle-eyed look of a cowboy. Just like the men, women ranchers had to have strength enough to operate machinery, handle a team, put up hay, work cattle, repair fences, clean ditches, and handle a horse well enough to herd cattle. Trail rides didn't always happen on pretty days with blue skies above and wildflowers below: it could be raining or snowing or a trial of heat and flies. But there was that "code of the West" tradition of being tough: a cowboy worked if he was sick or injured—he was never a quitter. There was a kind of pride in this toughness, and the women had to develop it, too. It was said of tiny, wiry Marie Scott that she could outwork many men.

Part of the appeal of ranching to many of these women came from their association with the ranchmen in the business. While most of the women in this book either had a short-term marriage or didn't attempt it, indicating that in the end they may have preferred their independence, there were hints here and there in every story of the enjoyment of the company of cowboys. The best of the men in the ranching business—the ones many of these women associated with—were the hard-driven and successful ranchers of their communities, and they were a pleasure to be around. They could be hard-headed and stubborn, it's true, but often they possessed a unique western gentlemanly kind of charm. Though they may have completed only an eighth-grade education—not uncommon in the early years— many were intelligent and well read—had become experts on such things as water law or genetic science—and were extremely interesting to talk to. They were savvy, confident, and knew exactly what they wanted in life and weren't afraid to go after it. Some collected small empires of land and cattle. They could be affectionate and easy with women. Although always gentlemen, the words "honey" and "sweetheart" came easily to their lips. But it was always said as a compliment, and never mind that they may have spoken the same words of affection to their favorite horse. Genevieve Hardig was said to have had strong friendships with the ranchmen in the Delta community. Marie Scott enjoyed her association with my grandfather, Clyde Buffington, whom she referred to as a "good head." She would often call him for advice and just for the sake of conversation, as he was an able storyteller.

With ranch women, there were no clear parameters where children were concerned. Some of the women wisely chose to remain childless, knowing that their ranching careers would take every bit of their energy and time and feeling that children would possibly suffer neglect as a result. The problem for these childless women was that while they may have managed to build empires, at life's end, there was no one to pass the ranch on to. There were even occasions when women were accused of neglecting their children in their great quest to make a ranch successful. Ironically, the ranch had to prosper in order for a woman to put food on the table for these same children. Some women were able to successfully combine the role of mother and ranch woman. This was sometimes made possible because of the good will of another woman—an aunt or grandmother—who would step in and help out. There was agreement among the women on one thing, however—combining children with ranching wasn't a problem that their male counterparts had to worry much about. Interestingly, more traditional ranch wives seemed to have great sympathy for the plight of their sisters who were attempting to run ranches and raise children; they admired and respected those who succeeded at it. Perhaps they knew more than anyone else the difficulty of doing justice to both jobs.

All of the women ranchers in this book seemed to have had a special relationship with the land. They would get to intimately know every rock and rill of their ranches and every season of the meadow. They would know about when to expect their favorite wildflowers to arrive—the very first yellow avalanche lilies in the aspen groves in late May or the fiery Indian paintbrush on the hill in June. They would know where the rare little fairy slipper orchids hid in the pines and how the irrigation water sounded coming down the hill—guessing a ditch's fullness just by its sound. The smell of the clover in the meadow, the song of the aspen leaves in the breeze, the way the shadows fall from the mountains across the land at evening time, the thunder crashing from mountain to mountain in a sudden summer rainstorm, the wail of a coyote at night—it all became a part of them. A woman involved with ranching had a mother-earth kinship associated with the birthing process of animals, too: the delivery of a calf or lamb could bring her close once again to the experience of bringing new life into the world.

There were women, then, who thrived on ranching. "Having what it takes" was a prerequisite and a characteristic of each of the nine women featured in this book. There may have been others of the same ilk who could have been included—each of them unique and interesting—but these nine were the ones that stood out as shining examples. The reign of the women of

Chapter One Having What It Takes

Riding, Roping, and Roses spans a time period from the late 1870s to the present and includes women from many parts of the state, both the Eastern and Western Slopes. These women, one and all, were worthy of admiration—true ranchers, businesswomen, and women of steel, yet womanly—some of them mothers, some wives. They were great role models for other women, well ahead of their times, and all proved—whether motivated by necessity or an appetite for a way of life—that they were equal to the very difficult task of ranching.

Zelma Nesbit at her family's ranch in 1944.
Photo courtesy of Zelma Nesbit Kruger.

Chapter Two

MARIE

Marie Scott
(1896-1979)
Alpine Meadows Ranch—Ridgway, Colorado

"Every morning I hit the floor at 4:30 ... I work 'cause I like to."

Marie Scott quote— *The Denver Post*, May 29, 1983

Marie Scott spent her final days on a couch in the kitchen of her simple frame ranch house. She died peacefully on November 5, 1979, not far from the home where she had been born eighty-three years earlier. The place where she was to spend her last days had been of her own choosing—she had hoped not to die in a hospital. There were no children or grandchildren or, for that matter, any other close family with her to say good-by—to ease the passing. None existed. But friends were there—friends who passed for family—a family whose bonds were forged not with blood, but with those strong-as-steel ties that existed in the old ranching communities of the West.

In the gaunt and wasted woman there was no hint of the girl that Marie once had been. The abundant curly red hair was now thin and gray. Porcelain skin had weathered into leather from too many days in the sun. Eyes once the color of mountain bluebells now were a clouded and washed-out blue. Oh, but there had been a time when she had turned many a young man's head—even though she might well have been able to outride, out shoot, and out cowboy even the best of them.

All that was forever in the past now, and the worn-out Marie—her face haggard, her thin legs bowed, her hands gnarled and worn with work—had decided to die in the comfort of her familiar, almost shabby kitchen with its ancient cook-stove, linoleum floor worn off under the sink, geraniums spindly in the window sill, and a beloved dog and a few treasured cats nearby. There wasn't really anything so unusual about any of this—except that almost everything you could see of the beautiful view from her window and more—she owned. Thousands and thousands of acres worth millions of dollars. And she had spent her lifetime and her life's blood earning it.

Marie may have died a legend and a very wealthy woman, but her birth more than eight decades earlier had been a common enough one. She certainly had not been born with a silver spoon in her mouth. Her father, Bartley Scott, had come from Sand Hill, Missouri, to the west slope of Colorado in 1882, along with his parents and many other hopefuls, to take part in the settlement of the town of Ouray, Colorado. In 1883, the family moved to the nearby Dallas Creek valley located some ten miles from Ouray in the heart of incredibly rugged and beautiful mountain ranching country. Her mother, Ida Josephine Culver, came to the same area around 1890 to fill a position as school marm in a one-room school in the fledgling Ridgway ranching community. A short teaching career was not unusual for the West's early-day schoolteachers, with marriage ending many of them. In 1893, Ida followed the school marm cowboy-marrying custom and wed rancher Bartley Scott, and they quickly began a family. But before marriage, Ida had accomplished something quite out of the ordinary for a woman of

Marie's mother, Ida Culver Scott, on the left and her grandmother on the right. Date unknown. Marie Scott photo, courtesy of Mario Zadra.

her time and position: she had purchased a large house—a home of her own that she tenaciously held on to through the years ahead and the trouble to come. Perhaps this mother with her stern Scottish and English heritage was an example early on to Marie and her older sister, Loraine, of just what a woman could have and keep if she set her mind to it.

Marie was born on March 28, 1896, on the family homestead, which was located in the Dallas Creek valley. All around this area, ranches and farms were springing up at this time. Her earliest experiences were of land and cattle and all that went with taking care of them. The seasons of ranching—spring calving, summer haying, fall roundups, and winter feeding—were ingrained in her psyche early on. The traditional male influence and protection of a father were not to be hers for long. Her father, from whom she had inherited her red hair and blue eyes, died at the age of thirty-four when Marie was only eight years old. The local newspaper, in the usual

J.H. Scott. Date unknown.
Marie Scott photo, courtesy of Mario Zadra.

flowery obituary of the day, stated that Bartley Scott was a man everyone loved and respected, a good husband and father, and that he had ailed two years with stomach trouble, dying after surgery to correct it. After his death, Marie's grandfather, J.H. Scott—who was a circuit-riding minister—would somewhat fill the masculine role in the girls' lives, but it was undoubtedly their mother who had the greatest influence on them.

The Ridgway community that the Scott family had settled in was similar to many other frontier towns on the west slope of Colorado in the early 1900s. It was a cow town frequented by local cowboys who tied their horses to hitching posts all along the main street. Ridgway was also headquarters for the Rio Grande Southern Railroad, and people, cattle, coal, and lumber were coming in and going out constantly on its rails. The huff and puff of the little engines with their big clouds of steam were an everyday sight. It was a rough little town with dirt streets and cowboy bars, but it was tempered by the influence of the churches and schools being built here and there along its narrow streets. Hard times came to the town in 1917, when many a young cowboy was taken from its population to furnish the manpower for World War I. Following on the heels of the war was the loss of many of Ridgway's residents to the Spanish Influenza of 1918-1920. Marie and Loraine experienced these events as a part of their frontier childhoods. Their larger community, the surrounding ranching society, was a fairly rough place, too. Outside the town was the usual collection of self-sufficient homestead ranches with their rough log cabins and barns, their huge gardens of vegetables and potatoes, and their hay fields, where the native hay grew lush. Civilization was coming slowly to Marie's childhood home, but it was still only a step removed from Indian territory and the grizzly bear and mountain lion still roamed nearby.

After Bartley Scott's death, Marie's mother continued to run the Scott ranch, developing and enlarging it and eventually putting a very young Marie in the role of managing it. Management also included cooking, which Marie was expected to do for the ranch's hired hands. Surprisingly, she enjoyed it, becoming an excellent cook. She baked up scrumptious cherry pies in the kitchen when she wasn't mowing hay in the hay field. A visiting uncle, John Culver, wrote in his hometown Decatur, Illinois, newspaper in an undated article that a teenage Marie was managing the entire large Scott ranch when he visited his sister there. He greatly admired his niece and

A young Marie Scott horseback in her earlier ranching days.
Marie Scott photo, courtesy of Mario Zadra.

described her as "wearing khaki and puttees, a sombrero and long gloves." He also said of her, "She says that she is never happier than when she is riding a horse that gets the notion that he is boss and tries to unload her. She can ride, shoot and declared to me that she had never seen a man who could pitch more hay than she can pitch."

Managing the family homestead wasn't enough to satisfy Marie for long. Accounts vary, but when she was sixteen to eighteen years old—and having already stated in no uncertain terms that after the eighth grade, she was through with school, telling her mother that she would rather dig ditches; and having already collected something of a herd of cattle as best a young child could . . . strays and sickly calves, true, but a start in the cattle

Marie in a large brimmed, high-crowned hat, jodhpurs, and high-laced boots—attire favored in some parts of the West in the early years. Circa 1920. Marie Scott photo, courtesy of Mario Zadra.

business—she next bought a place of her own. And so, before reaching the age of twenty, and with borrowed money, she owned her first small acreage: by the age of twenty-one, she was adding on to it.

By her thirties, Marie was going along about like everyone else in the ranching business, when the Depression hit the small Ridgway ranching community hard. Marie survived it by hunting deer and elk and by trapping and selling pelts. Many others did the same. Even with the hardship, the Depression turned out to be a time of opportunity for Marie. She had gone to work for the U.S. Land Bank, and she was sent by them to Kansas City on business. Before she left, she sold some of her cattle at a railroad shipping point at a town called Placerville, which was located just over Dallas Divide, several miles from Marie's ranch. Receiving part of her money for them there and the rest when they and she arrived at Denver's stockyards, Marie was flush with $6,000 or $7,000, which she intended to wire home to her bank in Ridgway. Her mother may have changed the course of Marie's life when, in a phone call, she relayed the news that the Ridgway bank had just folded. Marie was left in the enviable position of having a lot of Depression cash on her hands. Land prices were falling now, and some in the cattle business were having to sell out. Marie bought 3,000 cheap acres in Norwood, some thirty miles west of her home ranch. She was on her way.

It was about this same time that Marie did something quite out of character—on September 28, 1929, in Telluride, Colorado, she got married. She had always had admirers, but they soon discovered that pursuing a land- and cattle-driven woman on horseback and getting no encouragement was not worth the effort. Bob Valiant must have been something different. Tall compared to the five-foot Marie, straight in the saddle and with a reputation as a fine cowhand with a fine little herd of Herefords, he won her over. There were different theories as to why. Some said it was the Herefords. Some said that she had helped nurse him back to health after an injury and had fallen in love with him. Some said she felt pressured by the conventions of a conservative and tight-knit community, where marrying was just the expected thing to do. Others said it was easier for her to get banks to loan her money to buy more land as a married woman. Only Marie knew the reasons. But she wouldn't have been the first or the last to fall for the way some men look on a horse or the feel of strong, callused hands that know how to be gentle in private moments.

Chapter Two Marie Scott of Ridgway 25

A rare photo of Marie and Bob Valiant in a wedding party of friends. Marie, holding a cat, was rarely seen in a dress. Date unknown.

Marie Scott photo, courtesy of Mario Zadra.

Whatever Marie's reasons for hooking up with Bob Valiant, the marriage was probably doomed from the beginning. She was thirty-four years old, well used to her independence, and totally driven to complete her goals of buying more and more land, fencing it all in, and putting the finest Herefords behind the fences. Bob was of a similar strong-willed nature, and the two had some legendary battles. A final clash occurred while they were driving cattle, and it resulted in Bob being told "someone's going to leave here and it isn't going to be me." A divorce took place, which was final on July 26, 1937, with Marie citing extreme and repeated cruelty. She then bought out Bob Valiant's holdings, a thing that few women of that time would have been able to do. Then she took back her maiden name.

Marie never expressed much regret about her failed marriage—perhaps husbands and children didn't fit in with her formidable ambitions, and she was wise enough to know it. A few years later, Marie hired her former husband (who by then had a new and maybe more conventional wife) to work for her. She is credited with saying about Bob Valiant, "he wasn't much of a husband, but he was a hell of a good hired man." Marie rarely swore, so perhaps this statement is somewhat of an exaggeration. Another comment attributed to her on the subject of her former husband was, "I traded him off for twelve cats some time ago and have been the better off since." Marie had a wicked sense of humer and comments of hers, such as these, were quoted and repeated around the Ridgway community until they became legend. Marie, described by men who knew her well as a "hard twist" and "tough as a pine knot," may well have had such a view of a man who had once unsuccessfully tried to disagree with her. Even so, it was Marie who took care of Bob Valiant in his declining years, providing her former husband, down on his luck, with food and shelter.

After her divorce, Marie went right on with her one great quest—acquiring more and more land—or, as she put it, "outsmarting some 'old papas' on land deals." From Depression times on, Marie continued buying up valuable ranch land, often for defaulted taxes, pledging as collateral her own growing assets. It took nerves of steel, and Marie had them. Until her death in 1979, Marie's life followed pretty much this pattern. Her brother-in-law, Will Harney, in an article about the Scott sisters in a local recipe book, stated, "If a bucket of dirt was for sale, Marie would buy it."

Marie's "look" really never changed much either. She always dressed in Levi's and had the curious custom of some of the '40s- and '50s-era cattlemen of only wearing new ones instead of washing them and wearing them again. Of course, she wore boots and western shirts, favoring the color red for her shirts and jackets. Until she was elderly, she was a natural redhead or an unnatural one, when her hair began to gray. Generous amounts of red rouge—"war paint," she called it—was her trademark, as was the way she wore a Stetson-style hat well back on her head. She was short at five feet, slender, and more and more bow-legged as the horseback years added up. Marie was rarely seen in a dress, but she had her feminine side. She often had her hair done at a salon in Montrose, and she sometimes drank the coffee that kept her going from her mother's pretty china cups. She wore an apron over her Levi's when she cooked, and she could turn out as good a cherry pie or as flaky a biscuit as any housewife around. She had some colorful western lingo to spice up her conversations, and her word for "yes," for which she is well remembered, was always "yeah-yuh."

Marie in her kitchen in 1972, baking a cherry pie. Photo courtesy of Mario Zadra.

Just as Marie wore plain ranch work clothes—no beaded and fringed buckskin to impress upon people that she was a rich cowgirl—so was her manner of living simple. Marie's ranch house, the one she lived in much of her life, was located at the heart of her Alpine Meadows Ranch near Ridgway. It was a modest and plain frame house with red trim, surrounded by the usual outbuildings, corrals, sheds, and barns of an everyday ranch. There was not one ostentatious thing about it, considering it was long lived in by one who literally owned most of the valley where it was located. Other than its great size, there was nothing about Marie's ranch that advertised her formidable standing—it was indistinguishable from any other ranch in the area. Marie wanted it that way.

The ranch house was made homey with a wood- and coal-burning cookstove in a kitchen that brought comfort to her as well as many a house cat and ranch dog through the years. Marie often claimed that the more she knew of people, the better she liked her dogs. Marie's love for animals was obvious and hers lived like kings. Dogs accompanying Marie on her rounds in her Jeep were ordered hamburgers or steaks at restaurants. Cats were

Marie's house had a huge window in the living room with a magnificent view of the green Dallas Valley and the great stretches of grass that were her irrigated pastures.
Photo courtesy of Mario Zadra.

given leftover steak on newspapers spread on the kitchen floor, and if they wanted to nap on the kitchen table, then that was all right, too. The same round oak table also served as a desk/office and was often covered with piles of records that Marie, and only Marie, knew the logic of. Marie's empire eventually reached so far beyond the view from her window that it is simply hard to imagine. Her land continuously grew and shifted boundaries and changed hands to the point that it is hard to get a handle on what she owned in her lifetime. Generally speaking, she eventually owned land from her home ranch on Dallas Divide just west of Ridgway all the way west to the Utah state line and beyond. Marie claimed to have never counted it all up, but estimates of her holdings at the zenith reached from 60,000 to 100,000 acres. It was an extraordinary ranch comprised of vast stretches of hills and valleys, mountains, sagebrush flatlands, and green blooming meadowland. It was checkerboarded throughout with Bureau of Land Management and United States Forest Service lands and complex with water rights. Most of Marie's land was located in several counties in southwestern Colorado including the whole of the basin of Dallas Creek. She also owned land in the central Colorado Rockies. In addition, Marie's empire was studded with the gems of eight beautiful mountain camps. It was to these that her cattle and sheep were sent to graze during the summer months. It was a glorious ranch that Marie had built up—beautiful, bountiful country—splendid cattle country in its prime.

Owning these thousands of acres and miles of fences was one thing—managing it was something else. Compared with her vast land holdings, Marie usually owned relatively few cattle—around 300 head—which she kept at her home ranch. She bought her prize Hereford breeding stock from as far away as Texas, but purchased most of it in Colorado—from the Potter Hereford Ranch near Ouray, the Buffington Hereford Ranch and the Swietzer and Field Ranch near Gunnison, and the Redd Ranches in Utah. Her cattle were top notch, and she pampered them with diets rich in native hay, fortified with her feed mix of oats and other ingredients that she kept secret. Marie was caught up with the prestigious registered Hereford business, as were many others of her time. The Herefords, beautiful English cattle sometimes called "white faces," could do extremely well in high-altitude locations such as Ridgway. They could endure on short rations when necessary, survive extremely cold weather, mother good calves, and produce

Work was accomplished at Marie's ranch with the help of one or two hired hands and additional ones for haying season. Pictured here in 1962 is one of Marie's longtime and faithful hired men, John Howard.

Photo courtesy of Mario Zadra.

good beef at an early age. The Hereford breed was the cattle choice of the West, from the late 1800s until the 1950s and 1960s, and considerable "show ring promotion" added to its glamour. Ranchers showed up in coat and tie when looking for prospective bulls to turn out on the range, and spending big money for a pedigreed Hereford was a good way to impress one's neighbors.

The care of Marie's some 300 good registered cattle at the Ridgway ranch would have been a full time job for any cattleman, but she also had the rest of her immense acreage to manage. Fencing it required a year-round crew to rebuild and maintain some 300 miles of fences and gates. She hired help for other ranch jobs as well and expected a lot from her hired men. She joked that her hired help owed her the hours from 3:00 a.m. to 10:00 p.m.—the rest of the day was theirs to do with as they pleased. Marie talked constantly to her help, giving continual directions, which were sometimes resented.

In spring the irrigation ditches, bridges, and head gates came up for review and repair, and negotiations began with neighboring ranchers who leased much of Marie's land for the grazing of sheep and cattle. Spring was also calving time, and Marie's Hereford calves came into the world well-attended—often by her, personally. She kept track of each new offspring in a small black book—but mostly in her mind—memorizing the good and bad traits of each.

Calving season was hardly over when Marie needed to line up her strategy for mixing up land, leases, sheep, and cattle. This juggling act would have been simple on an average-sized ranch, but with Marie's great spread, it took some considerable doing. Added to this was the operation of one of the largest irrigation projects on the Western Slope of Colorado. Grass that started growing under her constant effort of irrigation was mature by late June, and then the real work began. The harvesting of hay required a full-time haying crew of at least fifteen men and a cook. While this was going on with Marie's hands-on help, a large fencing crew, hired and managed by Marie, was elsewhere hard at work. At the same time, the eight mountain camps that housed herders or riders that tended the cattle and sheep at higher elevations needed regular inspections—by Jeep or by horseback—by Marie. Little wonder that it was said of her, as reported in *The Farm Quarterly,* Fall 1960, "Marie Scott never sleeps—Marie Scott ain't human."

Winter work may have been a little easier on Marie, but Ridgway's weather didn't let up on anybody. Marie—a holdout for feeding hay with a team of horses—often faced temperatures that could drop to thirty degrees below zero or colder and four-foot depths of snow. In a corduroy cap with ear flaps, Marie braced for the cold and never let her prize Herefords miss a meal. Winter feeding was a grind for Marie and her hired help—a daily winter task. The process was one of harnessing a team, loading hay onto a sleigh, and getting it to hungry livestock by stringing it out to them, the cattle bawling and trailing along behind. The horses, pulling the sleigh and laboring with steamy breath, gave forth a warm, live, and powerful image—a picturesque scene in Ridgway's beautiful meadows under its towering mountains—but it wasn't all that romantic on the days when it seemed as cold as Siberia.

In winter or any other season—on any given day—Marie had a multitude of other tasks to consider. In addition to the ranch work she participated in, there was a regular schedule of banking, supply ordering, land buying, land selling, and land leasing to attend to personally. She hired no ranch manager, she had no computer—she carried the records of her multi-million dollar empire in a small black book and in her head. She put in tremendously long days and seldom took one off; vacations were unheard of. Her personal war with beavers was time-consuming as well, and Marie carried dynamite and blasting caps in the back of her Jeep, bouncing over all kinds of terrain. She practiced a philosophy of staying out of the beaver's territory, but only if they stayed out of hers. They usually didn't, and she spent much of her time staging dynamite warfare on beaver dams. About her beaver problems or anything else, Marie didn't ask for much advise—none from women and none from men who she considered inferior and who she called "old papas." The men who she trusted for advice were called "good heads," and she sometimes called upon them and carefully listened to what they had to say, and then generally did what she wanted to anyway.

For a long time, it seemed that nothing could stop Marie's great ambitions. But "Old Father Time" eventually came to call, and Marie, who had gone at a breakneck pace well into her sixties, was finally defeated by old age. She simply gradually wore out and could be seen noticeably limping or napping beside the road in her red Jeep. Her eyesight began to fail and, in her seventies, she broke a knee cap and was temporarily confined to a wheelchair—her great vitality slowly began to slip away.

At this time, Marie began selling much of her land, reducing her holdings to some 25,000 acres at the time of her death in what was undoubtedly an effort to spare her estate exorbitant death taxes. But it wasn't enough to cheat the government, whose regulations she had often battled against. The IRS collected seventy percent of the value of the last of her land—nearly seven million dollars—at the time of her death. The State of Colorado collected about a million dollars to cover state death taxes. What was left was divided among twelve heirs—Marie's friends and neighbors. Much of her remaining land had to be sold to permit equal division among her beneficiaries. Marie's great empire was broken now into pieces large and small and was never to be the same again. The character and the purpose of her beautiful Alpine Meadows Ranch and her additional holdings—all bearing the stamp of her strong personality and the benefit of her singleness of purpose—was as lost as grass visited by prairie fire.

Marie was buried not far from her home in a lovely small cemetery, the resting place for many of her life-long neighbors as well as many of her family. Under a pine tree, Marie Scott was put to rest in a wooden casket, wearing her western clothes—Levi's and a red jacket. She had requested only a simple grave-side service. She had asked that instead of flowers, donations

Marie Scott in her prime at Alpine Meadows Ranch. Circa 1970.
Lawrence O'Nan photograph. Courtesy of Ranching History of Ouray County project at Ridgway Public Library.

be made to a guide dog foundation. She left behind very few people who were close enough to her to know her well, and these, if they survive, still guard her privacy as she would have wished. Much has been written of her, often times repetitive—the "Calamity Jane" vignettes that have been told and retold so many times. What drove Marie Scott to such heights of ambition for acquiring land and to sacrifice all to that goal remains something of a mystery—and she would have wanted it that way.

Chapter Three

TWEET

Tweet Kimball
(1914-1999)
Cherokee Ranch — Castle Rock, Colorado

> *"I quit buying art and started buying bulls."*
>
> Tweet Kimball from the video recording,
> *A Jewel in the Rockies-The Story of Cherokee Ranch*

When Tweet Kimball unloaded her herd of thirty-one cherry-red Santa Gertrudis cattle—fresh from the famous King Ranch of Texas—at her Cherokee Ranch in Castle Rock, Colorado, some—including her ranch manager—said she was crazy. In fact, he refused to unload them, saying that this breed of cattle had never been and never could be raised in Colorado's cold climate. Tweet fired him on the spot and acquired some help from the house to unload the cattle. The "help" was a butler and the "house" was a veritable castle. Tweet, a very wealthy ranch woman of the fifties era, set out to prove her critics wrong and went on to successfully raise her exotic breed of cattle for nearly fifty years. She was fond of giving her bulls strong names such as "Cherokee Commander" or "Cherokee Governor,"—burying them along her driveway with bronze plaques to mark the spot when they died. Tweet was no ordinary, hardscrabble ranch woman; she had plenty of money to take the rough edges off that occupation.

Tweet was wellborn and beautiful to boot. A pampered only child and a true Southern belle, she received her education from the best schools for young ladies, was doted on by her distinguished father, and adored by her mother, who had been an Italian countess. She grew up in a genteel world with such amenities as old money, an elitist's education, fine art, literature from the libraries of the world, gracious manners, and, of course, servants to take care of the baser things in life. Early on, Tweet's father introduced his daughter to the world of high-class horse afficionados and trained her to be a skilled equestrian. On a visit to a Kentucky horse show with her father in the late 1930s, Tweet got her first glimpse of a slightly more rugged version of the animal breeding business—the registered cattle business—and a taste of the considerable "show ring glamour" that went along with it. The cattle she saw while at the horse show had come from the wide-open spaces of the West—from Texas—where life was just a little bit rawer than that found in the gentle Kentucky bluegrass country.

The cattle that so attracted Tweet were a breed called "Santa Gertrudis"—exciting to her, perhaps for their connection to the lore and history of Texas cattle ranching and, specifically, to the famous King Ranch, whose founder was the bold trailblazer, Richard King. In 1853 with his partner, Texas Ranger Gideon Lewis, King cheaply bought a vacant land grant of some 15,000 acres, called the "Santa Gertrudis," which was located in Nueces County in south Texas. King and Lewis commenced adding thousands of acres of adjacent property to their ranch, buying up stringy Longhorns from border towns and hiring skilled Mexican vaqueros to care for

A beautiful barefoot child holding flowers—Tweet Kimball in the 1920s.
Photo courtesy of the Cherokee Ranch and Castle Foundation.

them. Soon, Lewis and King's cattle numbered in the thousands and their huge empire continued to thrive and grow—in time the ranch included over a million acres. The King Ranch soon extended its operations to include the raising of blooded horses and, by the twenty-first century, would expand into oil, gas, and timber ventures, operating in several states and countries. When Tweet first became acquainted with the ranch in her early twenties, it was famous the world over and had been for some time.

The horse race and horse show businesses had probably given Tweet and her father their first contacts with the fine animals of the King Ranch. The ranch's breeding and racing of quarter horses and English thoroughbreds was well known—in time producing "Assault," who won horse racing's Triple Crown in 1946. It was about the same time of Assault's triumph on the race track that the King Ranch introduced to the world a new breed of cattle, one they had been developing since 1908—the Santa Gertrudis. By the 1950s Tweet, who had been so caught up in the romance of the horse world in the past, decided to move West and add the flavor of cattle ranching to her life. She had, for some unexplained reason, never forgotten the Santa Gertrudis she had seen earlier.

It seemed an unlikely match between Tweet and the Santa Gertrudis. These huge brutish looking cattle had been bred for the practical purpose of withstanding west Texas' hot climate. They weren't particularly pretty, by standards of cattle beauty, at the time Tweet went into the business in the fifties. They certainly could not be compared to the favored breed in the West during those years—the beautiful white-faced English cattle—the Herefords. The innovators at the King Ranch who had brought into being the new Santa Gertrudis breed had for years experimented with a cross between Indian Brahman and British Shorthorn cattle, and they finally established a breed that was five-eighths Shorthorn and three-eighths Brahman. In 1940, the U.S. Department of Agriculture recognized the Santa Gertrudis as a new breed of beef, the first developed in the United States. These huge animals, usually a deep reddish color, were significantly heavier than other breeds, had a high tolerance for hot weather, and were resistant to insects. An average-sized woman like Tweet, when standing at the shoulder of a Santa Gertrudis bull, stood level with the hump behind his horns. Below his horns were floppy ears and from his chest hung a huge brisket. But the Santa Gertrudis were rangy, well muscled cattle with a formidable

Chapter Three Tweet Kimball of Castle Rock

One of Tweet's magnificent Santa Gertrudis bulls.
Photo courtesy of the Cherokee Ranch and Castle Foundation.

beauty all their own, born of the sheer presence of uncompromising size, strength, and substance. They would be an impressive kind of cattle for Tweet to raise, and she would go about it in her accustomed way—high, wide, and handsome.

Tweet's wealth allowed her the luxury of buying a whole herd of blue-ribboned cattle to start out with. Most other ranchmen or women had to accomplish this the hard way, building their business up for years and paying their dues in blood, sweat, and tears along the way. But Tweet had clearly started out in life with advantages, having a much different background than her eventual contemporaries in the cattle business. She was born in Georgia in 1914 to Mildred Montague Kimball and Colonel Richard Kimball, who was a West Point career soldier and the eventual commandant of Fort Oglethorpe in Tennessee. The new baby was given the name Mildred Montague Genevieve, but her father referred to her affectionately as "Tweet" even before she was born, and the name stuck for the rest of her life.

Tweet grew up in Chattanooga, Tennessee, and was schooled at Pennsylvania's Shipley School and Bryn Mawr College. The influence Bryn Mawr had on Tweet was apparent in her attitude and in her accomplishments all through life. Beautiful, historic, graceful Bryn Mawr was a warm college home to Tweet—an ideal place for her to hone her intellect by participation in its rigorous academic program. Here, she would pursue her interests in the finer things, with studies in the history of art and architecture. Bryn Mawr probably never expected to produce a woman who would take on western cattle ranching as an occupation, but at least they had prepared her to do it with "class."

Whatever else the young ladies of Tweet's day were expected to accomplish, most assumed that marriage and motherhood would stand at the forefront. Tweet followed suit and married in 1938 while in her early twenties. An outstanding beauty to start with, Tweet made quite the picture of a bride in her demure white gown and a soft veil resembling a Spanish mantilla, which set off her green eyes and auburn hair. After the nuptials, a reception was held in the Kimball family mansion in Chattanooga, celebrating the union of Tweet to Merritt Ruddock, a prominent member of society who would, in time, become a member of the U.S. Diplomatic Corps. Tweet followed her husband abroad to London when he was assigned a member of the Foreign Service. As a young wife, she gamely played the same role as other well-married women of her time—collecting art, acquiring culture, and enjoying the company of well-connected friends. The Ruddocks added two adopted children to the family, sons Kirk and Richard, rounding out the picture of what must have appeared to be the ideal family life.

However favored by wealth and worldliness, there was apparently something amiss in the fairy-tale marriage of Tweet Kimball and Merritt Ruddock. After more than a decade of marriage, they were divorced. Tweet, being the classy woman that she was, never discussed this matter much. In the back of her mind, though, in spite of the pain of divorce, was still the dream of the West and the Santa Gertrudis cattle. Astounding her friends and family, Tweet—never one to be down on her luck for long—bid her Eastern upbringing and European wanderings good-bye, took her children, and traveled West, looking for a new start.

Tweet's divorce settlement was largely responsible for her move to Colorado, her ex-husband helping her buy property. Tweet always claimed that he had indicated that he would appreciate it if she located somewhere west of the Mississippi—far from their home in the South. Tweet took advantage of the offer, which must have been a generous one, and after looking around

Tweet Kimball and sons, Kirk and Richard, on horseback in 1955.
Photo courtesy of the Cherokee Ranch and Castle Foundation.

several western states, settled on a property near Castle Rock, Colorado, in 1954. Called "Charlford," it included a castle built by the former owners and 2,500 acres. The twenty-four room, fifteenth-century style castle, while hardly a typical ranch house, sat on land that was particularly well suited for ranching. She immediately renamed the place Cherokee Ranch. Tweet had picked some magnificent and rugged country on coming to Douglas County. The Rampart Range jutted out over the valley below at an altitude of close to 10,000 feet, forming the western boundary of the county and paralleling the beautiful South Platte River Canyon. Eventually, Tweet would purchase five more ranches in this area, allowing her to own, for a time, some 8,000 acres. It was country that had glorious history, too, having seen its share of Indians, trappers, explorers, and miners before settling down to become cattle country. In time, much of this wild and lovely country would give way to development and sprawl, a thing Tweet hated and did her best to prevent from happening to her beloved Cherokee Ranch.

Tweet's ranch and castle were located in the Plum Creek Valley, the castle perched on the top of one of the many high bluffs and mesas—perched so high that the view of the Front Range below was breathtaking. The grand old castle, built in the 1920s, was a tangle and wonder of gables, vaulted rooms, huge balconies, massive stone fireplaces, and even a Grand Hall. Tweet would spend the rest of her life here and entertain many prominent guests, among them Princess Anne of England. For such visitors as these, she would need the proper atmosphere, so she filled up the castle with the fine things of her youth and travels: Dresden china, valuable book collections, family antiques, and art. And with that taken care of, she then went about her main job—filling her land up with cattle—the best that money could buy.

From the start it was a battle. Colorado and local ranchers weren't exactly sending out the welcome wagon. They undoubtably viewed Tweet as nothing more than a wealthy "hobby rancher," and one not to be taken too seriously. Besides, Colorado was a Hereford stronghold. An innovative native rancher hardly dared to raise Black Angus on his ranch, let alone the foreign (from Texas, yet) Santa Gertrudis breed. Tweet wasn't too interested in what her contemporaries in the cattle business thought, only commenting that she thought them narrow-minded. She was a woman of definite opinions—strong-willed, demanding, hard-headed, and used to getting her own way. Beauty and wealth helped. But that aside, she was determined to find success in the cattle business and, eventually, she would do it in spades. Success was hard won, though—it would take years.

Tweet worked incessantly from the onset, starting out early in the morning on the telephone and continuing on through the day, taking care of ranch business. Of course she had hired help and, in fact, went through a succession of ranch hands and managers through the years. It was not these men, experienced though they may have been, but Tweet who made all the decisions about the ranch, down to the last detail. In time, she got pretty good at it. Although she never built fences, irrigated, or ran equipment like many other ranch women, she saw to it that it was all done her way. Carrying a clipboard, pencil, and binoculars, and possibly dressed in a business suit—sometimes even a mink coat—Tweet was out in the pens, amongst the cattle, and right in the thick of things. She determined at the outset that she would raise good breeding stock, producing the bulls whose good genetics would be passed on to the next generation. In the process, she intended to not only build up her herd size but her reputation as a serious cattle breeder, as well. She became adept at detailed record keeping and public relations, coming up with various campaigns to push the Santa Gertrudis breed. With patience and care, she studied genetics, took in sales, and designed advertising that would take her Cherokee Ranch cattle nationwide and, eventually, worldwide. By 1961, Tweet had established the Rocky Mountain Santa Gertrudis Association. There was no class for this breed at the National Western Stock Show in Denver when Tweet got started, so she helped establish one and, soon, she was winning its top prizes. By 1981 her cattle had won many ribbons—quite a few of them purple—with one being named Grand Champion Bull at the National Western.

At the zenith of Tweet's career, she was selling her cattle all over the world. She had even implemented a breeding program that produced calves year round to accommodate sales to ranches in any season. She shipped calves to Canada in the summer and to Texas in the winter. Her Santa Gertrudis stock also found homes in Taiwan, Australia, South America, and South Africa. Her cattle at Cherokee Ranch were well cared for— pampered—fed twice a day and down-right cherished by Tweet. She claimed that the selling points of her Santa Gertrudis were their production of a high percentage of calves, a high weight gain, and the ability to be weaned at heavier weights than other cattle. She looked them over often, driving through her pastures in a green Suburban and peering through her dusty windshield with big glasses on the end of her nose. As any good rancher does, Tweet knew her cattle by sight—remembered who sired or mothered each one and usually the names she had given them, and knew their weights, ages, and good and bad points.

In the beginning of her career of raising beef, Tweet seemed to have tried to dress the part of a cattle woman. Early pictures show her at livestock shows in the typical garb of the day—frontier pants, a white western shirt, and cowboy boots. Meticulous about her appearance, she always wore immaculate make-up and had her hair coifed to perfection; in later years, she dyed it to its earlier auburn color. She always looked classy, but truthfully, never very "western," even in cowgirl clothing. The effect of her efforts seemed to produce a strange combination of both her old and new worlds, her Eastern and Western lives. As she grew older, her tastes seemed to revert more to her origins, and she would be seen at sale barns— usually pretty earthy places—dressed to the nines, sometimes wearing her mink coat. When it got too hot, she would roll the mink up and place it under her seat at the sale barn or behind the seat of her car. As she aged, Tweet seemed to grow less and less concerned about her appearance and, according to some who knew her well, had little interest in anything but her cattle.

Tweet's life, aside from her cattle business, included an impressive list of civic achievements. She was a busy woman, involved in many organizations over time, many of them connected in some way to land or animals. She served for a period of time on the Planning Commission of Douglas County, was a trustee of the Denver Art Museum, held positions on Douglas County's Water Advisory Board and Educational Foundation, and chaired the Birds of Prey Foundation, among others. Tweet loved animals, as was evidenced by the care she gave her cattle. Horses were a big interest, too. Riding horseback, both English and Western, was a life-long pleasure and one she enjoyed until quite late in years. Dogs were another passion, and she always had one or more around—rumored to have the run of the house—or, in Tweet's case, the castle.

Along with raising her two sons and caring for her castle, cattle, horses, dogs, and a ranch teeming with wildlife, Tweet found time now and then to include a man in her life. But never for long. She married three times after her divorce from Merritt Ruddock—to an accountant, a military man, and a college professor. For whatever reasons, these marriages didn't last and, true to form, Tweet didn't discuss the subject much. She always had a full social schedule, however, and with her worldwide interests in the cattle business and her local and Denver civic responsibilities, she seldom had any time on her hands. Until the end of her life, however, Tweet's main occupation was raising Santa Gertrudis cattle. From this vocation she never retired, never failed, and never lost her drive and ambition.

In 1996, Tweet began a process that would enable her ranch to live on in the grand tradition that she had intended and fostered for forty some years. She established a private foundation, called the "Cherokee Ranch and Castle Foundation," and put her land under a conservation easement that ensured it would never be developed. Douglas County purchased the two million dollar easement, and the Cherokee Ranch and Castle Foundation held title to the property, now some 3,000 acres, which was considered to be worth, at

Chapter Three Tweet Kimball of Castle Rock 45

Tweet Kimball: a pretty woman, a fine horse, beautiful purebred cattle, and magnificent country to ride through.
Photo courtesy of the Cherokee Ranch and Castle Foundation.

market value, many millions of dollars. Tweet's unbelievably generous donation of this land allowed it to become permanent open space—a sanctuary for wildlife and the means to preserve the exquisite views that surrounded the castle. A board of directors and staff would run the foundation and the castle would be opened for public tours. But most importantly to Tweet, the Santa Gertrudis would still be raised at Cherokee Ranch.

A castle perched atop a hill is a fine place to live, but for one living alone, it was undoubtably lonely towards the end. Considered more and more eccentric as she grew older, Tweet appeared to some to be an intimidating and power-hungry elderly woman. She was a paradox of personality traits, however, and those who knew her best described her differently—as a polished, articulate, passionate, caring, and considerate woman. All agreed, however, that she was a true cattle woman and a true rancher, one with great pride in what she had accomplished. Grande dame Tweet died in 1999 at the age of eighty-four of a heart attack. Of all the accolades she received in numerous newspaper articles at the time of her death—much of it concerning her castle and its contents, her civic contributions, or her donation of land to the public—a tribute in the January 20, 1999 issue of *The Weekly News and Castle Rock Chronicle*, by veterinarian Matt Myles, probably would have pleased Tweet the most. He described her as "a cattleman in the grandest tradition and history of that title."

Chapter Four

GENEVIEVE

Genevieve Hardig (1884-1974)
Payne's Siding Sheep Ranch—Delta, Colorado

> *"Sometimes I think I'm strange as Dick's hatband."*
>
> Genevieve Hardig's use of an old time saying to describe herself, quoted from Bob Barns of Delta

---

Miss Genevieve, as she was often referred to, didn't look much like a cowgirl or a rancher. She always had the appearance of a blue-blooded lady—one from an earlier era, when ladies wore modest knee-length dresses, hats, gloves, and pearls. She dressed like those ladies of yesteryear who appeared matronly in their twenties and grandmotherly by their forties. Miss Genevieve—not quite five feet tall—was quiet, modest, sweet natured, and—in her younger years—even pretty. She looked every inch the part of a doctor's or minister's wife—one who served tea cakes at church suppers or headed up the local Red Cross polio drives back in the '50s. However, Miss Genevieve never was anybody's wife, and she didn't have much time for charity work, either. She was too busy running her businesses; she owned an abstract office where she worked long hours, and she put in equally long hours managing her sheep ranches.

Miss Genevieve's lifelong career as a sheep rancher took place in the small town of Delta on Colorado's west slope, some forty miles south of Grand Junction. No existing records or dimming memories shed much light on how she happened to come to this community in the early 1900s at the age of twenty-two. There is little mention of her in local newspapers— perhaps because she was a very private person and one who mingled very little with local society. Her main socializing seemed to be as a devout member of the Catholic Church. As she grew older, her reputation hardened a little, and she was then more likely to be described as severe, determined, strong-willed, resolute—a woman well ahead of her time, a shrewd business-woman—quiet and not particularly sociable.

Chapter Four Genevieve Hardig of Delta

Miss Genevieve, an only child, was born May 10, 1884, in Denver, Colorado, to Lothar and Mary Elizabeth Hardig. Apparently from a prominent family, she was educated at Denver's Loretta Heights Women's College and the American Business College of Pueblo. She perhaps joined family who had moved to Delta, arriving from the Eastern Slope shortly after her graduation. In the early twentieth century, the town of Delta was just moving out of its rough-and-ready frontier period. The old boardwalks on Main Street were being replaced, and the town had an electric light plant, and a fairly posh hotel—the Delta House. Medical facilities now existed to care for injuries, childbirth, and the common diseases of the day—typhoid, diphtheria, and influenza. Churches and schools had already been established, and the ladies of the town were collecting funds for the building of such things as libraries, hoping to add some culture to their frontier cowtown.

Upon Miss Genevieve's arrival in Delta in 1906, college degree in hand, she went to work for County Judge M.R. Welch. The Judge, a manly, handsome man some twenty years Miss Genevieve's senior, had

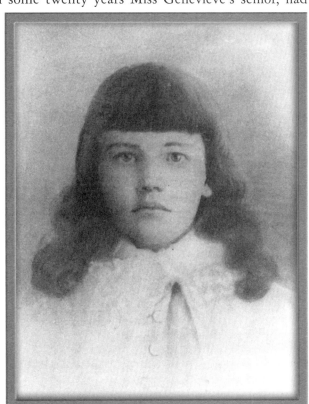

Genevieve Hardig as a young child.
Photo courtesy of Delta County Historical Society.

been engaged in farming in the Midwest before his entry into law school. According to local sources, these two were business associates for a time, becoming involved in several ventures together. By 1911, when Genevieve was twenty-seven, and possibly in some kind of partnership with her father or the Judge, she became an owner and operator of Delta's abstract office. In these young days of the century, and herself very young, Genevieve began a lifelong habit of acquiring land and then—seemingly totally out of character—turning it into sheep ranches. Perhaps the Judge, with his agricultural background, had encouraged her. It was brave of her, though, for Miss Genevieve's sheep ranches were located deep in the heart of cattle country.

Delta's cattlemen, in the late 1800s, had laid claim to a big handsome valley with the wide Gunnison River running through the heart of it. Formerly the home of Ute Indians, who wintered in the area because of its mild climate, the valley was soon occupied mainly by white men. First came the beaver trappers, who traded at Antoine Roubideau's Fort Uncompahgre on the Gunnison, and later the miners and homesteading cattlemen. Located between two great plateaus—the Uncompahgre and Grand Mesa—the Delta area was situated within easy reach of plentiful grass, and, in the early days, *free* grass. And so, it was ideal cattle country. Grand Mesa, the largest flattop mountain in the world, was coveted by the early-day cattlemen. Its high mesas were covered with great reaches of rich native grass that were sustained by the abundant water of five rivers and the 300-some lakes on the mountain's summit.

Early on, Grand Mesa was populated by thousands of cattle—grazing unregulated—their numbers increasing yearly. But trouble soon came into this new and self-designated "cattle country"—unwelcome interlopers—the "large predatory sheep outfits," as the cattlemen called them, from Utah. By 1890, the Utah sheepmen were bringing in large bands of sheep whose numbers ranged from 10,000 to 40,000, according to local reports, and the animals were wiping out the cattlemen's grass, even as they traveled through it. They carried with them the possibility of putting the cattlemen, who depended on free range, out of business. The sheepmen were met by armed and angry masked cattlemen on horseback, who shot and clubbed their sheep with green aspen clubs, and sometimes ran them over cliffs. Similar incidents were happening in many parts of the West.

Chapter Four Genevieve Hardig of Delta

The sheep and cattle situation was still tense by the time Miss Genevieve decided to go into the business. In fact, it stayed tense through years of trouble until it was resolved by the government, when a division of range for sheep and cattle was established. In 1905, the United States Forest Service took control of the higher ranges, and in 1934, the Taylor Grazing Act put the lower unregulated ranges under federal control and the feud ended. Ill will towards sheep and their owners, however, didn't end. Early local homesteader and rancher Enos Hotchkiss had kept sheep early on as a sideline to his cattle ranch, a practice that was at least tolerated by the cattlemen. This was probably the case with Miss Genevieve as well, because the numbers of her sheep at first didn't reach the threatening point. Ignoring the cattlemen's generally low opinion of sheep ranchers, she kept right on building up her sheep business and, over time, turned it into a fairly large operation.

Miss Genevieve bought her first sheep ranch around the time of the Great Depression, just as the sheep and cattle war's bitterest conflicts were ebbing. She may have admired, as many women did, the mystique of the ranchmen of the Old West—plentiful enough on the west slope of Colorado at this time. Many women of her era married into the supposedly "romantic" lifestyle of ranching. After marriage, these women took on the traditional role of a rancher's wife—the domestic role of performing housework, raising children, and occasionally assisting with the chores of the ranch. Miss Genevieve, however, disregarded tradition and set about creating the lifestyle of western ranching for herself. Time and again, she bought up land and sheep and hired a herder—usually a Mexican national or a Basque from northern Spain—to care for the sheep, then "looked after the operation," as she put it, never getting her hands very dirty in the process.

Because of her abstract office business, Miss Genevieve had a handle on available land before and during the Depression, and she took advantage of this knowledge. She was an astute businesswoman and her sheep business was quickly becoming more profitable because during World War I, the Army needed wool for military uniforms and blankets, which caused sheep prices to climb higher and higher. Miss Genevieve probably also found sheep raising economical, because bands of sheep could be partially pastured on public lands for relatively low fees. When not grazing on public lands, sheep could live on alfalfa and other cheap feed—such as the refuse from the harvest of the sugar beets, which were grown in the area. And while prices for market lambs and wool declined during the Depression, America's entry into World War II brought about another demand for wool, and prices climbed again, bringing Miss Genevieve even greater profit.

Raising sheep was proving to be practical for Miss Genevieve in other ways. Because of their flocking tendency, only one herder and two well-trained sheep dogs could handle a band of a thousand sheep, so hired help could be stretched pretty far. Of course, there were pitfalls in Miss Genevieve's business, too. Sheep could be poisoned by milkweed, which was very toxic in the spring, just as cattle could be poisoned by the beautiful blue-flowered larkspur plant. And predators were a constant threat. In the early days, grizzly bears were common, as were wolves, mountain lions, and coyotes, and any of them could kill several sheep during a night's visit. Once Miss Genevieve learned how to deal with these problems and caught on to the business—learned how to ride out the cycle of good and bad years—her sheep raising changed from a questionable venture to a lifelong commitment.

Judging from pictures of Miss Genevieve's sheep taken in the '60s, she raised a mixture of several good breeds, such as Hampshire and Suffolk, profiting from two cash crops every year of wool and meat. Lambing was what began the sheep raiser's seasons of work. Lambs came the first of March and, for three weeks or so, a close watch had to be kept on the flock. By April, as the early grass was greening, Miss Genevieve and most local sheep ranchers were holding their sheep bands on their own grass at home ranches. But by mid-June the herders, with a thousand pairs of sheep each, began either trailing or trucking animals up to summer range on Grand Mesa, keeping close watch over them until their return in mid-September. Sheep ranchers had winter grazing permits on the nearby Uncompahgre Plateau, which was used in the winter months, with sheep coming back to home pastures in March for shearing. Miss Genevieve had to hire more help—up to ten or twelve men who could shear a thousand sheep a day.

Even though she left much of the day-to-day operation to her long-time herder, known as "Dirty Jim" because of his reluctance to bathe, Miss Genevieve oversaw the business herself on a regular basis. She likely ran into some problems trying to run this large operation while working full-time—and doing it all as a woman in a man's world. Hired help wasn't always to be trusted, and the unreliable type might have been tempted to take advantage of an "ignorant" woman. She gradually earned a reputation

Chapter Four Genevieve Hardig of Delta

as one who could hold her ground, thus receiving the respect of her contemporaries—the local sheep and cattle ranchers—and eventually even buying a few of them out.

Profit was not the only reason Miss Genevieve found sheep ranching an enjoyable business. The fever of the great outdoors—the freedom and independence that went along with mountain ranching—would have been a direct contrast to the long hours she spent pouring over recordings in her abstract office. Miss Genevieve and the other ranchers of the Delta area certainly had beautiful surroundings for their life's work. At the foot of Grand Mesa, where the sheep ranches generally were located, was country called the "dobies," which were comprised of clay bluffs, domes, and hills of a silver-gray color. The dobies presented a barren landscape—looking almost like the surface of the moon—but had an unusual kind of beauty. Some of Miss Genevieve's home ranches were located near such country.

Since ranches in or near the dobies had very sparse grass, sheep ranchers needed summer grazing permits on Grand Mesa to augment their animals' feed. This huge mountain jutting up into the sky behind Miss Genevieve's ranches had plenty of grass. Its terrain graduated from the dobies at its base, up slopes of sage and higher to the juniper, pinon, and scrub oak, then on to the top, where the abundant aspen, spruce, and fir trees grew. Miss Genevieve's views when on the upper slopes of Grand Mesa were spectacular—in summer, even more so. Then, the flat top of the mountain was carpeted with wildflowers amongst a multitude of blue lakes.

While Miss Genevieve was running her sheep ranches and working in the abstract office—often for twelve hours a days—she also created a lucrative sideline of what were known as her "land deals." Many were the parcels around Delta that she bought, traded, or sold throughout her long life. Forty-five acres of land northwest of Delta, where she was growing sugar beets, was sold in 1945 to become a subdivision. Like Ridgway's rancher Marie Scott, Genevieve became adept at locating, buying, and then selling land for a later profit and had the nerves of steel that were required to pull it off. In fact, she and Marie were fast friends and occasionally compared notes in the confines of Miss Genevieve's abstract office—one dressed in western clothes, boots, and Stetson, and the other in very proper feminine attire—but identical in their formidable ambitions.

Genevieve Hardig on the left, in her Delta abstract office, circa 1927.
Photo courtesy of Delta County Historical Society. Ben Walker, photographer.

Although Miss Genevieve owned ranches and other property all over the county, she preferred being in town. For most of her life, she lived in a substantial looking yellow brick house in the 800 block of Delta's Main Street near her abstract office. This large home, with its several dormer windows jutting out in all directions, was a cozy place with vines climbing up the front walls, wide front steps, and the shade of large cottonwoods. Today, the house looks as sturdy as when she last lived there. Miss Genevieve lived alone in this house that she filled with antiques, but one could surmise that she mostly just slept there, with her business ventures claiming her time for long days and many weekends. For a time, Miss Genevieve also owned a cabin high on the slopes of Grand Mesa near Bear Lake. She spent some time there in the summer months and—as was her custom—much of it alone.

Miss Genevieve's largest ranch—her main ranch—was located near Highway 92, northeast of Delta. It was purchased by her in the '40s, and she owned it for many years, considering its acquisition the pinnacle of her long career in the business of buying, selling, and trading land. Her property was located near a railroad loading point known as Payne's Siding, and while the ranch may not have had any official name, locals called it "Miss Genevieve's." Her ranch consisted of some two thousand acres in the dobies. Miss Genevieve tried to spruce the place up some and improve it by creating a couple of lakes, but even then it was barren looking and not a very pretty sight. It was prettied up some when she had a Spanish-style house built for her longime, trusted herder, Dirty Jim. But looks aside, it was still good sheep country. Off in the distance, though, one could see the great vistas of the Elk Mountains and the Raggeds Wilderness in one direction, and in the other, Grand Mesa towering over the valley below. Sadly, nothing much is left now of Miss Genevieve's great sheep ranch, just some falling-down corrals; the house, and even the trees she tried to grow near it, are gone.

Miss Genevieve lived out her long life in the small community of Delta, working and working, buying and selling, never having the time or possibly the interest in having a man or children in her life. She often described herself as "strange as Dick's hat band," blaming it on being an

Genevieve Hardig at her Payne's Siding sheep ranch in the 1960s, looking across a band of her sheep to the house she had built for her herder.
Photo courtesy of Delta County Historical Society. Ben Walker, photographer.

only child, but she went right on being unconventional until the end. She always seemed happy enough, though, according to her friends, and she was generous at Christmas with her customary gifts of fruit cakes to one and all. She turned down her chance once at being Marshall of the Deltarado Days parade, lining up some other of the "old fathers" of the community to take the honors. She believed herself to be still too young to be in that position. She didn't think of herself as old and never did, even though she was the same age as her replacements in the parade. She insisted on driving her black Chrysler through the town and country, even though, when she was elderly, the other residents began to steer clear of her for safety's sake. Her style of dress, though long out of vogue, never changed, either. To the end, she wore her ladylike dresses, favoring the color black, and although a bonafide rancher, she rarely—if ever in her life—was seen in the boots and jeans that were a trademark of that occupation.

Miss Genevieve retired from her abstract business after forty-nine years and lived on in her Main Street yellow brick house, attending St. Michael's Catholic Church and socializing with the Business and Professional Women's organization occasionally. But she mostly ran her sheep ranches. She lived to be ninety years old, having devoted most of her life to work and a few close friends. After her funeral at St. Michael's, where she had requested six women pallbearers, she went home at last to where she had come from some seventy years earlier and was buried at Mount Olivet Cemetery in Denver. Miss Genevieve's godchild, who was named after her; a few distant relatives, mainly cousins, and most living in other states; and St. Michael's Catholic Church were her beneficiaries. Her estate was valued at $150,000 at her death, not including the value of a considerable amount of real estate. A street in Delta bears the name Hardig Drive, but thirty years after her death, there are probably only a few in town who have enough memory of her to know the significance of this. For the most part, Miss Genevieve seems to have been nearly forgotten, as often happens when there are no descendants to carry on a family's heritage.

Those few who do remember her speak of her with curiosity—often with a chuckle—but always with respect for this tiny lady, who, armed with intelligence, ambition, and a strong will, paved her own way in that most masculine of all men's worlds—the cowboy's world—and did it all in white gloves and pearls.

Chapter Five

MABEL

Mabel Steele Wright
(1898-1993)
Wright Ranch — Creede, Colorado

"These things I love: currant bushes in the sun, columbines in the rain, little brown eyed boys, and this ranch on the Rio Grande..."

from *River Ripples*, by Mabel Steele Wright

Mabel Steele Wright spent her entire fortunate life high in the San Juan Mountains of Colorado enjoying some of the world's grandest scenery and "living the dream" that so many have dreamed—that of the cowboy way of life. She was born on November 1, 1898, at the ranch home of her parents—Charles and Cynthia Steele—Lake City pioneers. Lake City was a thriving mining camp at the time of the birth of this family's third child. Gold fever had earlier brought the first population of hopefuls to Mabel's birthplace—long and narrow Hinsdale County. Lake City was high country, surrounded by five spectacular 14,000-foot peaks of the San Juan Mountain range. It was comprised of steep terrain, deep canyons, and rushing rivers. The narrow river bottoms grew lush native hay and provided the location for the first homesteads in the valley. Mabel spent her growing up years in Lake City and the remainder of her long life only a day's ride away in Creede, Colorado—an equally beautiful place situated in the same spectacular mountain range. Through marriage, Mabel was able to make a transition from one lovely mountain ranch to another—her new ranch home located close to yet another boisterous mining camp—Creede. The silver and gold bonanza camps of Lake City and its sister settlement Creede did not, like many others in the Rockies, fade into history as ghost towns. Cattle ranching ended up sustaining many of the new hopefuls, and following fast on its heels was tourism. Families like Mabel's were able to stay on when gold was not to be found or when silver prices plummeted—by taking their chances instead with land and cattle.

Chapter Five Mabel Steele Wright of Creede

Mabel Steele's father, Charles, was a native of New York and came West when he was little more than a boy to try his hand at cowboying in Texas. In the early 1890s, he moved up north to Colorado and the Lake City area to work for the Denver and Rio Grande Railroad as a tie contractor. Three years after his arrival, he married Cynthia Weed and settled on a homestead ranch a few miles north of town. He then set to work supplementing his railroad income with the profits from a small herd of cattle and a large garden. Tragedy, no stranger to the early settlers, first struck the Steele family when two of their children, little pioneers Lee and Dolly Steele, died in 1901 of a fatal childhood disease. Then, in 1916, Charles suffered a stroke that left him paralyzed and bedridden for the rest of his life, eventually resulting in his early death. At the time of his death, the family numbered six children. Cynthia carried on with raising them as best she could, and with a remarkably cheerful outlook. She undoubtably served as an example to her daughters of how a woman could carry on alone— even manage a ranch on her own. Mabel witnessed a very strong pioneering work ethic in her mother, and it influenced her greatly later in life. Cynthia,

Charles Steele, Mabel's father, as a young man.
Kraig Saunders photo, courtesy of *Lake City Silver World.*

Cynthia Steele, Mabel's mother, on her wedding day, January 1, 1893.
Photo courtesy of *Lake City Silver World.*

during her husband's illness and after his death, ran the family ranch almost single-handedly. The Alta Vista Ranch on the Lake Fork River continued to thrive under her care, supporting cattle and chickens and a huge garden. She created much of the family's income by producing cheese and butter from the milk and selling the produce from her garden to nearby Lake City residents. She was a cheerful woman, and in later life, wrote of her difficult and sometimes hair-raising experiences on the ranch with calm acceptance, a sense of humor, and no hint of self-pity.

Young Mabel was a typical ranch girl, and by the time she was four years old, she was riding a pony and taking responsibility for part of the family's chores. The oldest surviving child of the family of six, she also helped in raising her younger brothers and sisters. Said to be her father's favorite, Mabel was grown and leaving home by the time he died. During his long illness, she was a mainstay to him and also to her mother, taking

The Steele's ranch home north of Lake City.
Photo courtesy of *Lake City Silver World*.

Chapter Five Mabel Steele Wright of Creede

School in the Steele family living room with furniture removed to make way for desks.

Cynthia Steele photo, courtesy of *Lake City Silver World.*

on a fair share of responsibility for one so young. She and her siblings were schooled at home—their classroom was the homestead cabin's living room, and the teacher lived with the family. When she reached high school age, Mabel boarded in Lake City during the week. Riding a horse to school was probably the only other alternative, but that was too much of an ordeal during Lake City's frigid winters. The simple social events of the day served as her entertainment: church ice cream socials, spelling bees, and country schoolhouse dances, where a pretty young girl might find a cowboy to two-step, waltz, or square dance with. She was, indeed, an attractive young lady during her high school days; slender with long dark hair falling in curls down her back, and smart as well, graduating as class valedictorian. Shortly after graduation, Mabel passed an examination that qualified her, at the tender age of seventeen, to teach school.

Teaching credentials in hand and hardly dry behind the ears, Mabel left home on horseback on her mare, Firefly, shortly after high school graduation and began her career as a teacher—a career that didn't last too long. Easing her emotional good-bye from her family was a contingent of several neighbors and friends who rode with Mabel and Firefly over Slumgullion Pass, southeast of Lake City, escorting her on to her teaching job at Hermit Lakes—a high mountain ranch that raised both cattle and fish. Here, she boarded with a ranch family and taught summer school for eight students from three neighboring families. She adapted quickly to her new life, even assisting with the ranch's commercial fishing venture. She recalled in her memoirs, "From then on I was too busy, too interested, too utterly fascinated and happy for anything else." Mabel finished her first year of teaching at Hermit Lakes and then moved on to a "winter school" located nearby at the L.S. Officer Ranch, where she boarded with the Clayton Wetherill family.

Many an early-day schoolmarm's career was ended by marriage to a cowboy and Mabel proved to be no exception. A tall, stout, and fine-looking man, Ray Wright, son of a local rancher, courted the pretty school-

Mabel Steele teaching at Hermit Lakes.
Kraig Saunders photo, courtesy of *Lake City Silver World.*

Chapter Five Mabel Steele Wright of Creede

marm briefly, then married her in Creede on May 6, 1917, and moved her to his family's ranch on the Upper Rio Grande. Ray Wright was a typical hard-working, no-nonsense cowboy. He was a well-liked and pleasant man, but a man of strong convictions, and one who suffered no fools. A trespassing fisherman on the Rio Grande, which ran the length of the Wright Ranch, might find himself facing a horseback

Ray Wright in his mid-thirties.
Photo courtesy of Sue Knowles.

Ray, lassoed by his rope, then firmly removed from the property. Ray and Mabel—both of them confident, strong-willed, and ambitious people—were well matched and soon forged a strong and lasting matrimonial bond.

Right off, married life seemed exciting to the new Mrs. Wright. For starters, Mabel's new hometown of Creede was interesting—

Mabel Steele Wright in her early twenties.
Photo courtesy of Sue Knowles.

considered one of the most notorious of Colorado's early mining camps. It had been immortalized by a famous line of poetry written by Cy Warman, editor of an early-day local newspaper:

> "It's day all day in the day-time
> And there is no night in Creede."

Creede had earlier been home to some 10,000 people, among them such notables as Calamity Jane, lawman Bat Masterson, and the famous card shark, Poker Alice Tubbs. It was also heartbreakingly beautiful, located near what is now the La Garita and Weminuche Wilderness Areas, and surrounded by alpine valleys and run through with clear mountain streams. Creede was one of the most productive silver mining districts in the state and also produced its share of gold, lead, and zinc. After the boom days and following a drop in the price of silver, some of the town's early population turned to cattle ranching. Mabel's new family had been early ranchers to the area, settling on the Upper Rio Grande in one of the valley's most pristine locations.

The Wright Ranch was comprised of an upper and lower ranch, divided by a neighbor's land that was sandwiched in between. Mabel and Ray lived the first year of their married life on the lower ranch in what Mabel described as a "honeymoon cabin," then moved to a big nine room ranch house on the upper ranch. This house, which survives to this day, was built around the turn of the century. It was a large, warm typical ranch house with a kitchen big enough to feed large haying crews, six bedrooms, and a cozy enclosed front porch. Huge cottonwood trees sheltered the yard and, off in the distance, an ever changing-view of the beautiful San Juans could be enjoyed. Mabel fit right in here—feeling at home—and, in fact, made the ranch her year-round residence for the next fifty years. After that, she spent her winters in Creede, but she returned every summer to the ranch.

Relatively few ranches existed in the Creede area, with the Wright Ranch being one of the largest, eventually growing in size to some 2,600 acres. It was a magnificent spread with great productive hay meadows on either side of the Rio Grande River. The ranch had permits on the nearby Rio Grande National Forest, which increased its grazing capacity. Summering cattle were brought into the Creede mountains regularly from the San Luis

Chapter Five Mabel Steele Wright of Creede

Mabel Wright's ranch house.

Photo courtesy of *Lake City Silver World.*

Valley, and the cattlemen or "riders" who looked after them were among the colorful people who came into Mabel's life in the early days. Eight years after their marriage, in 1925, Ray, Mabel, and two of Ray's brothers bought out their father's ownership of the ranch. They continued to raise Hereford cattle and prodigious amounts of hay, which took most of the summer to harvest. They often didn't sell their calves in the fall as most ranchers did, having enough hay to hold them over the winter and selling at a higher weight in the spring. Soon, they added to their well-established cattle enterprise a lucrative tourist business, building guest cabins to rent to summer visitors. For many years, the care of guests and cabins became Mabel's and her sister-in-law's job—cooking three meals a day, doing laundry, and cleaning cabins. World War II's food shortages ended the serving of three meals a day, but the cleaning and laundry work grew as more cabins were added over the years. The first cabins were primitive and had ice boxes and coal stoves but no running water. In spite of the roughness of it, the business was a huge success and people returned summer after summer, enjoying the scenery, the atmosphere of the ranch, and the chance to congregate at Mabel's big ranch house, where they played cards on the porch, sang around the piano, and enjoyed the company of genuine cowboys.

From the time of her marriage in 1917 until Ray Wright's death in 1955, Mabel enjoyed an almost idyllic life as a rancher's wife—a life that she described in her book *River Ripples*. Remembering the simple joys of that first summer of marriage at the honeymoon cabin, Mabel spoke of a garden and a strawberry bed she and Ray planted together, saying, "We were inordinately proud of ours [garden], for though the bed was not extensive, we had all the berries we wanted . . . We had strawberries and cream, and even strawberries and ice cream, just any ol' time." Later, in her book she describes an "elk spook" she participated in with her husband, dressing warmly enough to help chase elk away from hay stacks in snow that measured forty-two inches deep: "First, I don over my cotton anklets a pair of white wool athletic socks . . . Next comes a pair of felt bedroom shoes and my rubber galoshes with four buckles—the warmest footwear combination I have ever tried . . . Then I stand on the kitchen table while Ray ties my jeans around my ankles, over the galoshes, with a stout cord. My heavy, lined leather jacket zipped up, a pink wool scarf on my head, and lamb's wool-lined mittens, and we are off!" One has only to read between the lines here to imagine a contented woman and a happy marriage—a partnership of ranch adventures shared together. A later entry in her book bears this out further: "All's well! The light from the kitchen windows beams a welcome. We are home! . . . The leaves have gone from the aspen, except in a few places; and yet their going seems not to have detracted from the beauty of the surrounding hills waiting so quietly in the sunshine and shadows. But now I hear sounds; cattle bawling and the voices of men bringing the last of the Clear Creek Pool stock over the ridge north."

And so it went for Ray and Mabel for close to forty years. They were childless but had a rich full life, filled to the brim with good neighbors, good cattle, colorful hired men, haying, cabin cleaning, summer guests, fine horses, good friends, awesome scenery, and, of course, each other. They were partners in all this—best friends. Unfortunately, it didn't last long enough; in 1955, Ray—still a young and vital man—died of a brain tumor. Mabel wrote of this loss in one simple sentence in her diary: "Today I lost my beloved Ray." Although broken-hearted, Mabel didn't wear her heart on her sleeve. She picked up the pieces as best she could, as her mother had before her, and resolved to go on with the life she loved on the Upper Rio Grande. She took stock and stiffened her spine: she had a ranch to run.

Mabel lived to be ninety-four years old; the last thirty-eight years of her life, she was pretty much on her own. She now owned a good part of the Wright Ranch, and she never once considered leaving it for an easier life somewhere else. After Ray's death, Mabel and the two remaining Wright brothers, Warren and Wallace, ran the ranch together—circumstances that one might consider ripe for disaster, especially since the brothers didn't always get along. But the situation was made simple by some kind of unspoken agreement that Mabel was the boss. It was widely acknowledged that she was the decision maker when it came to running the ranch—there was nothing going on there that she wasn't involved in. She kept the books, looked after the guest cabins, and cooked for hired cowboys—and while the Wright men may have looked after the business of cattle ranching, it wasn't without Mabel's advice and consent. This arrangement with the brothers lasted until 1963, when Warren Wright was bought out by Mabel and Wallace. These two ran the business together for a few years until his death in 1969, and that event left Mabel in full possession of the ranch. As sole owner, the first thing she had to do was sell part of the property to pay inheritance taxes. It was a tall order for Mabel to take on the entire responsibility of a huge ranch that had once needed three owners to keep it going. In the early '70s, she brought her brother, Charles Steele, into the business, agreeing to let him buy a half interest in the ranch over a twenty-year period.

Even though Mabel was always involved with brother-in-laws or brothers in some fashion regarding the management of the ranch, she continued to be the mainstay. She "bossed" and she "juggled," and she saw to it that the ranch survived over the years. Both the cattle and the tourist businesses were kept going—her tourist business always thriving and her cattle business experiencing the traditional cycle of lean and prosperous years. Mabel did a staggering amount of work to accomplish this success, and she relished every minute of it. Along with the work, there was always lots of company at the ranch. There were many summer guests, visiting friends, and the hired cowboys. Mabel functioned at the center of this little society and was well loved by all—praised for her fabulous cooking and for her salty stories, which she told with a great deal of flare. She claimed that she was never lonely at the ranch at anytime. For her there was no other spot like it on earth.

The cattle ranching part of the Wright Ranch was pretty traditional—the usual mountain ranching practices were followed. In the early years, cattle were fed in the winter with a team and sleigh, usually a long winter's activity because of Creede's nearly annual deep snows. Spring brought calving season followed by fence repairing, irrigating, and harvesting hay. The huge ranch had to employ many hired hands to keep things in working order. The Wrights always had good cattle, keeping a large herd of commercial Herefords. That the ranch was a thriving, successful operation was no doubt aided by the summer tourist business.

Cowboys who hired on at the Wright Ranch spoke of their experiences there with affection, claiming it was a good place to work, even though ten-hour days were the norm. It helped that they ate "like kings," thanks to Mabel's cooking. She could have given all this ranch work up for easy street. More than once she was offered millions of dollars for the ranch, but she never seriously considered that option. People also offered good money for the right to hunt on the Wright Ranch—but Mabel didn't want their "blood" money, as she called it, and therefore she didn't allow hunting.

Mabel's cabin renting business was not an unusual practice on the Rio Grande. In fact, most of the ranches along the river had a number of primitive cabins to rent out to fishermen. The fishing was superb here with plentiful cut throat and native trout, weighing up to one or two pounds each. They were pulled out of the stream one after another; cowboys even fished for them on horseback. After a particularly prize catch, Mabel's guest fisherman would draw an outline of his or her fish on a piece of white paper, which would then be tacked up on the enclosed front porch-walls. Eventually, the entire porch was papered with these fish outlines, some dating as far back as sixty years. It was little wonder that guests returned year after year to experience the Rio Grande's great bounty of trout.

Changes came to the Wright Ranch when the cattle herd was sold in the early '60s. From the homesteading days on, cattle ranching had been the main purpose of the ranch, and this purpose was now retained by leasing pasture to ranchers from the nearby San Luis Valley. This move eased up the ranch work but not the care and upkeep of the ranch itself—its fences, ditches, and meadows. But, of course, the tourist business still thrived, and Mabel continued running it—she never really stopped, cleaning rental cabins until she was nearly ninety years old.

As Mabel aged, the care of the ranch became increasingly difficult during the winter months. In her middle seventies, she purchased a small home in Creede, closing up the big ranch house in the winter months and moving closer to civilization to occupy her time with local senior citizen's activities. She was on Creede's Soil Conservation Board and belonged to a few woman's organizations, such as Eastern Star. She was a well-known person about town—wiry and thin, with large glasses and casual clothes and sporting a no-nonsense look. She never really considered herself a senior citizen, but she would join the others for bus trips. One suspected that she was really just passing the time until she could return to her beloved ranch for the summer months and the tourist season.

Mabel died in July of 1993, leaving her part of the ranch to several nieces and nephews. Unfortunately, it had to be sold in order to pay estate taxes. The 800-acre upper ranch—all that was left of the original huge Wright Ranch—remained in the ownership of her brother, Charles. At his death, the property passed into the capable hands of his widow, Dorothy. She is currently in the process of preserving as much of it as possible with a conservation easement, which among other things, restricts its use to agricultural and preserves the beautiful old house and barn.

Mabel Steele Wright was one of those fortunate people who lived a long and happy life, claiming near the end of it that she was content with what she had accomplished. Ranch life had allowed her a certain freedom and a closeness with nature and beauty that was enviable. One could say her life was charmed, but it really wasn't. It was Mabel herself—through her good judgment, wise decisions, and her willingness to take on any challenge—that made for that charmed life. She had early on displayed courage when leaving home at a tender age to tackle the career of teacher in a one-room school. She had been wise in her choice of a mate and steadfast in her loyalty to him. Courage surfaced again when Mabel persevered in keeping her ranch and lifestyle intact when her world was turned upside down by Ray's early death. Mabel lived her life to the fullest, never remarrying but finding the need for family fulfilled through brothers, sisters, nieces, and nephews. When she died, she was buried next to her husband in a little cemetery in the town of Monte Vista, which is located southeast of Creede in the San Luis Valley. During her nine decades, Mabel had created a legacy that literally brimmed over with purpose—the people, work, mountains, meadows, and seasons of a beautiful ranch.

Chapter Six
MAVIS

Mavis Caldwell Peavy
(1904-2000)
The Rock Ranch — Logan County, Colorado

> *"Let me tell you something about the integrity of Mavis Peavy. If she said that she had a flock of chickens that could pull a wagon, then I'd hitch 'em up."*
>
> Auctioneer Jim Wingate at the closeout auction of the Peavy horses, *Greeley Tribune*, May 22, 1984

Mavis Caldwell Peavy spent the better part of her life struggling with the difficult job of ranching on her own. Beleaguered though she was with the heavy responsibilities of ranching—a job she had never planned to take on—she nevertheless made quite a success of it. But she probably would have given it all back for just one more day with Marshall Peavy, the handsome rancher she never quite got over—the one who had stolen her heart when she was young. Mavis first laid eyes on him when, as a young schoolteacher, she had let the students of her one-room school outside for a few minutes to take in the excitement of a passing cattle drive. Trailing a large herd of cattle from the Hahn's Peak area of Routt County was Marshall Peavy, who—after eying the pretty young schoolmarm—stopped for a few minutes to chat with her and the youngsters in her charge. Mavis regarded this tall, dark stranger—blue-eyed and handsome, spurs a jinglin'—from her perch on the schoolhouse step and promptly fell in love. Apparently Marshall felt the same, and the chance meeting—in time—resulted in marriage. It would be a love match tragically cut short by Marshall's early death; it would be the first, last, and only love for Mavis.

When Mavis's parents, Andrew and Anna Jane Caldwell, brought their daughter into the world in 1904, they must have had serious doubts about her chances. Mavis weighed in at an extremely delicate two pounds, and there weren't many medical miracles around at that time to up the chances for premature babies. But with a combination of good luck, good care, and a fighting spirit, Mavis survived, proving to be a trooper from the outset. When Mavis was three years old, her family moved from their home in Seattle, Washington, to Denver. The only home in this young girl's memory then was to be the mountains and plains of Colorado.

In Denver, she grew up with a typical "city girl" background and graduated from East High School in 1922. Young Mavis then took the standard teacher's exam of the day and straightaway accepted a job that took her to the mountains and to Moon Hill School—a one-room rural school located north of Steamboat Springs in Routt County. Mavis, with a hankering for the high country, found this teaching job had brought her to one of the most attractive mountain settings in the state. She decided to stay on there as a teacher and, in her brief career, taught at several of Routt County's rural schools—all in beautiful locations, with one situated at the bottom of Rabbit Ears Pass.

At the time of Mavis's tenure as a teacher in Routt County, the area had developed into a flourishing agricultural region. Ranches had sprung up all along the banks of the Yampa River and its tributaries—ranches of sufficient acreage to keep this beautiful valley open and undeveloped, forming perhaps one of the handsomest ranching valleys in all the West. Along with the ranching country, a culture had developed—one characterized by rough-and-ready cowboys, blooded horses, fine purebred cattle, a crusty, cavalier manner of doing and speaking, and a freedom and independence that was fresh as a mountain breeze and heady as homemade dandelion wine. It was a brand new world for Mavis Caldwell—halcyon days—and when they were interrupted by Marshall Peavy, those days became even better. The lively excitement—the warmth and companionship of this cattleman—brought to Mavis complete and ongoing happiness.

The object of Mavis's affection—Marshall Peavy—had come from Alabama to the West with his family, a mother and four brothers, around 1910. The young family located in Boulder and there, Marshall took to the mountains as if he had been born to them. In fact, he was so taken with the idea of mountains and horses that he left home at the age of twelve with a 30.30 rifle, a camera, and a packhorse, and spent a couple of years cowboying in Utah. At the age of fourteen, he and a brother rode their horses from Boulder to Steamboat Springs, and right then and there, he decided that the Yampa Valley was the place he was meant to be. With his mother's help, he and his brother eventually bought a ranch there and began raising Hereford cattle and ranch horses, as well as polo ponies. Marshall's first love would always be horses and in time he branched into the establishment of what would become a widely known quarter horse breeding program. When he married Mavis in 1925, it was a fine cow pony she received as a wedding gift—one of his favorites—one that could work cattle and win races.

For thirty years, Marshall, and eventually Mavis, built up a ranch in the Deep Creek area of Routt County, specializing in raising Hereford cattle and fine horses. As the years went by, more and more of the couple's attention went into the horse breeding program. They had started this business with horses of the famed Steel Dust line, whose progeny would eventually become widely known throughout the West and contribute to the foundation bloodlines of today's quarter horses. By the late 1930s, Marshall and two

Mavis Peavy on the right on Shiek, two neighbors on the left. Circa 1925 at the Peavy Ranch.

Photo courtesy of Jennie Mai Bonham.

Western Slope friends—Si Dawson and Coke Roberds—along with horse breeders from several other western states, gave birth to the new American Quarter Horse Association. Marshall ripened during these years into a man who looked every inch the part of a western gentleman. Weathered, and perhaps looking a little older than his age, pictures at this time show him well-dressed, prosperous looking, and wearing a fine Stetson that shaded his steely blue eyes and chiseled face.

At the time of his marriage to Mavis, Marshall probably expected that she would take on the role of a traditional ranch wife. And this role Mavis accepted, as would have been expected of any woman of her time. But she also took part in the business of breeding horses and cattle, especially the record keeping part, and enjoyed many good hours on horseback, often riding with Marshall. Motherhood took up more and more of Mavis's time as the years went by. Three children arrived in the family—a son who died in infancy and two daughters. The two girls, Mary Claire and Frances Jane, became Mavis's main responsibility, although horses and horse breeding were still an integral part of her life. People who knew her during this period of her life describe her as being a contented and happy woman, and one who was still practically mesmerized by her husband. She wore her good fortune well, though it was said by some that she had a certain aloofness about her—a "noble air." Those who knew her best, however, claimed that this was only window dressing and that underneath this facade, she was actually an unpretentious, warm, and friendly woman.

When their daughters were young teenagers, Marshall and Mavis made a decision to sell their Routt County ranch and buy one located at a lower elevation, perhaps finding ranching in mountain climates a little harsh as they grew into middle age. They picked out the plains of eastern Colorado as the place to relocate their ranching operation and, in December of 1943, moved the family to the Kenesaw Valley of Logan County. Their horses and cattle arrived by the train-car load at Buchanan, south of the town of Peetz. The sixteen carloads of cattle and three carloads of horses found their new pastures located at a place called the "Rock Ranch." The new place wasn't nearly as spectacular as the mountain ranch they had left behind, but it was nevertheless very well suited for their purpose of raising horses and cattle.

The Peavys were settling now in prairie country—Colorado's eastern plains—where wagon trains had once wound along the South Platte River and great herds of buffalo had roamed, where wildflowers and grass still carpeted the plains after rain, and the Rocky Mountains began to come into view, just barely visible in the distance. Here, ranches had first sprung up after the Civil War, locating near stage stations along the Platte Trail. Following these first ranches were the "railroad ranches" that would supply beef to the track-laying crews of the Union Pacific. The Peavy's Rock Ranch was in the heart of land often called "big sky country," where in some places one could see for fifty miles in any direction. There were great reaches of grass and the distant horizons hosted beautiful flaming prairie sunsets. The plains had always been great ranching country. Old evidence told the tale of the first ones who had settled there—the old post office towns and one-room schoolhouses of the turn of the century—standing lonely now out on the plains, many to disappear entirely in time. The prairie the Peavys came to was no longer home to the Indian, the buffalo, or the pioneer, but was now just as it always had been—home to wind and dust storms, blizzards, prairie fires, drought, and rattlesnakes.

As early as 1867, an enterprising man named J.W. Iliff had made his start as a cattle baron in this country, by following the early progress of the railroad crews along the Platte River. He had started up a string of ranches along the way of the railroad, buying thousands of longhorns to stock them. Others along the railroad route were doing the same. Eventually, the LF Cattle Company that Iliff operated would spread all the way to the Nebraska line. The ranch that the Peavys had purchased in this area was a unique one that once had been part of Iliff's great spread. Their historic Rock Ranch sat on the plains at the site of a good running spring and near a large cliff. At the time it was purchased in the 1940s by the Peavys, it was graced by windmills and many unusual and impressive rock structures. Barns (one huge one measuring 100 x 30 feet), numerous ranch buildings, houses, and even fences were constructed of rock that came from nearby. With its unusual rock architecture and rich history, the ranch—located on the Weld-Logan county line northwest of Sterling, would now be home to Mavis and Marshall Peavy.

The new ranch, the endless prairie, and the departure from beloved mountain country must have been quite an adjustment for Mavis Peavy and her husband. But the new surroundings *were* ideal for ranching. Mavis had her family to care for, and the horse and cattle breeding business was going on as strong as ever. Mavis was undoubtably still getting settled and trying to get used to all the newness of her situation, when her life abruptly changed forever. Marshall, roping a calf on the prairie on his favorite horse, Monte—in the prime of his life and with many productive years ahead of him—found the danger that has tripped up many a cowboy. His trusted horse stepped into a badger hole, causing an accident—a wreck for horse and rider that cost Marshall Peavy his life in 1944 at the age of forty-six.

For quite some time, Mavis literally floundered in grief and shock. When she lost Marshall, she lost her footing and found herself adrift in a sea of grief, loneliness, and uncertainty. Suddenly, and with no warning, she was alone with an 11,000-acre ranch to manage, and two teenage daughters to raise on her own. Worst of all, her confidence was shaken by the many nay-sayers amongst the ranching community who were quick to predict her failure. But . . . there were cattle and horses to tend to—especially horses—and Mavis slowly began to pick up the pieces of her life, take stock, and go on alone. She struggled from the beginning. Her granddaughter, Jennie Mai Bonham, described the effort: "In the wake of her husband's death, she struggled for the respect which would allow the latitude to operate a ranch as a woman in a man's world. My grandmother had to make decisions and judge character often under duress. Unfortunately, she occasionally fell prey unsuspectingly to a 'fleecing' by someone she let into her world with far too much trust."

As her life was changing form, Mavis physically changed at this time, too, and she who had been slender, pretty, and dark haired when young, at her husband's death grew into the stoutness of middle age. Now she left romance behind and determined to carry on with taking care of her ranch, her girls, and the livestock that provided their livelihood. In time, she adjusted and made a good life for herself. But she never stopped talking about Marshall Peavy until the day she died.

Soon after Marshall's death, Mavis had to let her job as housewife take second place. Now she would check on cows, inspect windmills, buy and sell cattle, raise good horses, and raise two daughters on the side. She tried to hire good men to help her, and—in this—she was sometimes disappointed. So, whenever possible, she called on her two young girls to help, knowing that their father had taught them much about ranching before he died. She and the girls often worked long hours—they faltered and made mistakes, but they picked themselves up and went on.

At some indefinable point in time it became apparent that Mavis was going to make it: her belief in "stick-to-itiveness" began to pay off. She was a bonafide rancher now, and she wore a Stetson-style hat and dressed western. She had no time for involvement in traditional women's social gatherings, or for much of anything else, for that matter, except running the Rock Ranch. Over time, she emerged as a happy, satisfied woman—one who had come into her own as a woman rancher and who was, as well, a loving and giving mother and grandmother. Eventually she became somewhat of a legend, some even calling her "Lady of the Plains." It is quite possible that it had been the horses that saved her.

Mavis Peavy in the 1960s.
Photo courtesy of Jennie Mai Bonham.

After Mavis took control of the ranch, it was the quarter horse breeding program established by her husband that became her new purpose in life. Marshall's breeding program, recognized as one of the top in the country, continued on under Mavis's ever more skillful care. She had a top-notch ranch to raise the horses on, and the Peavy reputation was already well established. But that didn't mean it was ever going to be easy.

Following in Marshall's footsteps was a very tall order. His renowned family-of-champion quarter horses previously had won top honors: "Saladin" winning the title of grand champion stallion at the 1938 inaugural National Western Palomino show, and "Gold Heels" taking the purple ribbon as grand champion stallion at the 1944 inaugural National Western Quarter Horse show.

Mavis continued with this successful breeding program and, with some experience behind her, decided to branch off into a new line. She began

Chapter Six Mavis Peavy of Padroni—Logan County 79

Mavis Peavy with her Herefords at the Denver Stockyards.
                                  Photo courtesy of Marsh Witte.

breeding Appaloosa horses and eventually earned her own blue ribbons. In 1988, two of her Appaloosa stallions were inducted into the Appaloosa Hall of Fame, along with Mavis herself—the only woman to be so honored at the time. The Hall honored her as a pioneer breeder who had raised several Appaloosa champions during her lifetime. By the end of her career, Mavis would be credited with many great contributions to the early-day quarter horse and palomino registries, as well.

Raising horses was an impressive occupation, one with prestige and connections to people who sometimes appeared to be as high-bred as the horses they raised. There was an atmosphere of sophistication and a certain flair to raising blooded horses. Horse shows were competitive, colorful, and high-flung events. That was the public side of the business, and it could be downright fun. The private side of the business at home could be even more satisfying, providing an authentic purpose in life—something close to genuine comfort and solace.

Observing the new foal lying in the straw in the stable, one saw the result of careful planning and breeding—and it was *alive*—its characteristics always a surprise . . .often a good one. The colts frisking out in the pasture were a satisfaction to view, too. One might even see in one or two of them the promise of the makings of a champion. Over the years, the gratification of her work increased as Mavis saw the strong Peavy bloodlines started long ago by Marshall appear in horses owned by people all around Colorado and other western states. Mavis found her salvation in raising horses—an occupation that would bring her meaning, joy, and recognition for the rest of her life.

Although Mavis accomplished much of her success in the horse breeding business on her own, she eventually made a decision to "partner up" with someone who could share some of the responsibility. Mavis never remarried, but after she had gotten firmly on her feet and her two daughters had married and left home, she did find a man to assist her as an employee and, eventually, as a business partner.

Ned Snader, as a young man, had come West from Pennsylvania to attend college at Colorado State University. His ambition was to become a veterinarian, but after being bumped out of classes by more advanced students, he ended up needing a job and was employed at the Rock Ranch. This young man, some twenty years younger than Mavis, was sized up by her as someone upon whom she might be able to depend. Snader was hired and a lifelong friendship began, which in time became a full-fledged partnership. Ned became Mavis's right-hand man, and the two of them continued running the Rock Ranch and breeding horses for many years. In 1968, this property was sold, and the two partners moved on to a ranch in northeast Weld County. They remained in business there until their dispersal sale in 1984, after which Mavis retired from ranching.

The dispersal sale, where almost the entire famed Peavy horse herd was auctioned off in May of 1984, was a huge event. Fence posts all along the route to the ranch were marked with banners to show the way. A huge orange and blue tent greeted the thousand or so visitors, many of whom had come from out of state. Barbecue was served to the crowd and then folks strolled to the stalls, where forty-three head of horses could be looked over by prospective buyers.

The auctioneer started off the sale by reviewing the history of the bloodlines of the Peavy horses. His review of national and international champion blood assured buyers they were getting their money's worth. Mavis was reluctant to see her horses go, but kept out a few of the older horses—"old partners" she called them—to keep on the ranch. These aside, this day ended her long and successful career in the horse breeding business.

Although Mavis spent her life engrossed in horses, she didn't neglect her family. She was a staunch supporter of her daughters and grandchildren; a loving and caring mother and grandmother who financed more than one grandchild's college education. She was a social woman, but mostly in terms of horse-related activities. She was a lifetime member of the American Quarter Horse Association and a member of the Appaloosa Horse Club, the Weld County Livestock Association, and the Cowbelles—an arm of the local cattlemen's association that promoted beef.

One of Mavis's greatest satisfactions in life was her involvement in 4-H. Every year, beginning in 1970, she sponsored a horsemanship award and trophy, which was presented at the Logan County Fair. Mavis believed that working with horses was beneficial to children, teaching them to think for themselves and helping them develop patience. Those who won her horsemanship award were given a large sterling silver engraved bowl as a trophy.

Mavis Peavy lived to the ripe old age of ninety-six, outliving her husband by more than fifty years. She was busy and productive until the end. She was survived by a daughter, seven grandchildren, seven great-grandchildren, and her long time associate, Edward R. "Ned" Snader III. From near and far and all around the community of New Raymer, where Mavis and Ned had retired, ranchers, horse people, and friends came to say good-bye to this woman who had lived so long and so well, had survived her critics, and had persevered—even thrived—at being a woman rancher. Granddaughter Jennie Mai Bonham's words serve as a fitting testimonial:

"Mavis Peavy placed her family first and helped her daughters and grandchildren every chance she could. She carried on thoughtful, insightful, and witty conversation often filled with an abundance of warm laughter. Her mind was sharp. Mavis was a school teacher in her early years and she always hungered to learn more, this brought a richness to her life until the very end. Perhaps Mavis was not born with ranching in her blood, but she infused ranching into her veins through the hardships and blessings she endured in a long and treasured ranching legacy."

Mavis Peavy in her eighties.
Rick Schulte, photographer—*Sterling Journal Advocate*.
Photo courtesy of Jennie Mai Bonham.

Chapter Seven

EMMA, ANNA, AND LIZZY LIZZIE AND EMMA

Emma, Anna, and Lizzy Becker
Lizzie and Emma Hansen
*Cousins born in the 1870s on the Becker Ranch
and the River Ranch—San Luis Valley, Colorado*

> *"Women did do (some) outdoor work in the cattle business, particularly if there were no men around."*
>
> Becker photograph caption—Library of Congress Cowboy Exhibit of 1983

The Hansen and Becker cousins came of age around the turn of the century on ranches located in the San Luis Valley in south central Colorado. Emma, Anna, and Lizzy Becker and Emma and Lizzie Hansen were some of Colorado's very first women ranchers. Their identical names cause a certain amount of confusion—now writing about them, and probably a century ago, also. These cousins were genuine ranch women (or girls) who branded cattle, rode the range, and did other ranch work. They performed these ranch duties—normally relegated to men— because they had to and also, it seems, because they wanted to.

When photographer O.T. Davis took the prize-winning photograph on the next page, he captured the Becker sisters branding in the old-time way. A calf was stretched out in a log pole corral, a small smoking fire nearby heated branding irons, and the sagebrush hills in the distance provided the background. Emma Becker was pulling the front feet of the calf taut while Lizzie stretched out the back feet and Anna wielded the branding iron. Many such branding scenes were caught by camera in the early West, and this one is very typical—except that the customary "cowboys" are "cowgirls." And while the Becker sisters did indeed brand cattle, one suspects that this particular picture may have been posed for the photographer. Calves generally were roped from horseback and thrown to the ground before being branded, and the girls would have had to accomplish this in their long high-necked dresses. The bonnets and high-button shoes they were wearing seem a little dressy for branding. Yet, here are Emma, Anna, and Lizzy Becker, defying the mores of the day and perhaps even common sense, dressed fit to kill, and branding calves.

Chapter Seven The Becker and Hansen Cousins of San Luis

The Becker sisters, left to right, Emma (14), Anna (22), and Lizzy (21), branding a calf on the Becker Ranch. This photograph was taken by a well-known turn-of-the-century photographer, O.T. Davis, in 1894; was displayed at the Library of Congress Cowboy Exhibit in 1983, and has also appeared in several other books and magazines.

Photo courtesy of the Monte Vista Historical Society.

The Becker sisters and their first cousins, the Hansen sisters—who lived and worked on a nearby ranch—were all teenage girls or young women at the time of the Davis photographs. Their time in the sun as ranch women or cowgirls was relatively brief. Information on their later lives is limited to mostly family recollections, which conclude that for most of the girls, after their fleeting stints as cowgirls, life was fairly commonplace. But for a time at least, they were authentic ranch women—their early-day cowgirl exploits a brand new and exciting chapter in the settlement of the West. The Becker and Hansen cousins, little known except for their famous pictures, definitely deserve their place in the roll call of Colorado's ranch women, yet their stories have never been told.

Anna, Emma, and Lizzy Becker and Emma and Lizzie Hansen were first cousins. Their mothers, Augusta Tessendorff Becker and Emilie Tessendorff Hansen, were sisters born in Germany in the mid 1800s. They

Emma Becker in 1894, riding side-saddle on a bull named John—cartridge belt around her trim waist—"six-shooter" holstered at her side.
O.T. Davis photo courtesy of the Monte Vista Historical Society.

immigrated in 1869 to the United States with their parents and siblings by way of the ship *Teutonia*. A year later, they came west by train and covered wagon with a large colony of German immigrants to the Wet Mountain Valley in Colorado Territory.

The Tessendorff girls, although quite young, were considered to be of marriageable age when they arrived in the West, where women oftentimes were a scarce commodity. Barely a year had passed when news of their presence got around, and nearby cowboy bachelors came a courtin'. Peter Hansen and his partner, Frederick Becker, came to look over the new arrivals from their recently established homesteads on an old Spanish property located on the Conejos River. Traveling by wagon, the two crossed the sagebrush desert floor of the San Luis Valley and, turning north, passed by Mt. Blanca and dropped over Mosca Pass into the Wet Mountain Valley. They found the two buxom, blue-eyed, German girls to their liking and, after making only a couple of visits—but seeming to think this sufficient—proposed marriage. The Tessendorff girls apparently judged these two cowboys to be satisfactory. Though older by more than a decade and hardened by the West, Hansen and Becker appeared to be gentlemen with good prospects. Their proposals were accepted and marriages soon followed, both taking place in December of 1871.

Traveling to their two new homes by wagon, the sisters—only fifteen and sixteen years old—arrived to rough and primitive dwellings of log and adobe, undoubtably little different than the ones they had left behind, and began married life to two hard-edged cattlemen who were little more than strangers to them. It took a lot of grit on both sides to take on such arrangements, and yet marriages such as these weren't uncommon in that day and time in the West. The Becker and Hansen unions proved in the end to be long-lasting and happy ones, perhaps strengthened by time and necessity. When the Hansens were older, Emilie would often refer to her husband affectionately as "you darned old mule," using the German word for mule. They were married for forty-two years. Since the Hansens and Beckers were neighbors, the sisters always remained close—so close, in fact, that they gave their daughters, roughly the same ages, identical names. At one time, there were even three girls in each family with names identical to their cousins, and this would have remained the case had not Anna, the third Hansen sister, died in infancy.

It was quite the valley—the San Luis—that the Becker and Hansen girls were born into—hardly fitting one's idea of a valley at all because of its great size. One of the largest mountain desert valleys in the world, the San Luis Valley reaches an altitude of over 7,000 feet and is roughly 100 miles long and 60 miles wide. Bordered by the Sangre de Cristo Mountain range and Mt. Blanca to the east and the rugged San Juans to the west, it is a place of long vistas and immense skies. The girls while in their cowgirl years, knew this place mainly as cattle country, although it had only relatively recently been so. This great, quiet valley had been home to several Indian tribes, and to the early Spanish priests. It had seen the passing by of explorers Coronado and Rivera, and later Zebulon Pike and John Fremont. Not far from the Becker and Hansen homesteads had been two of the West's most notable forts—Bent's Fort near LaJunta and Fort Garland near Alamosa. The forts' histories were first-hand knowledge to the girls' fathers, who had come so early into this country that they had seen much of its violence—had even known the likes of the Indian fighter and scout, Kit Carson.

Peter Hansen and Frederick Becker likely told campfire stories to their children, recounting their part in the settlement of the San Luis Valley when there were yet no towns and Colorado was still a territory and of their celebration when it became a state in 1876. These two ranchmen were already busy building up the country when the towns of Alamosa and nearby Monte Vista were born, where such early-day renegades as cowboy Rattlesnake Jack and his friends shot up the town and where the locals tried their best to outlaw liquor. The ranchers, along with other citizens of the new town, congregated to celebrate the arrival, in July 1878, of the narrow gauge train to the town of Alamosa. The coming of the rails spelled good fortune for them: now ranching could make sense—homesteads could be enlarged and cattle shipped out to new markets. The San Luis Valley, by the time the girls were born, was becoming superior cattle ranching and farming country, its semiarid climate enhanced by massive irrigation water—good and plentiful—originating from two major aquifers.

For the Hansen and Becker girls, the hardest work of ranching had already been accomplished by their fathers and mothers, who had carved out the early-day homesteads from nothing, and had led a harsh and grinding existence to do it. Frederick "Fritz" Becker, a large, powerful, and muscular man of a kindly and generous nature, had ventured west as a young man and had came to know intimately the rugged Sangre de Cristo and San Juan country. Later, he did a stint as a scout at Fort Garland, where he was firsthand acquainted with its war-toughened Civil War veterans. By also becoming acquainted with early-day Mexican vaqueros and Indians, he was able to gain even greater knowledge of his new country and its culture. He saw cattle raising as his greatest chance in this land of rough opportunity and set about building up a ranch with great determination. The pioneer spirit burned strong in Fritz Becker, allowing him in time to go from wandering mountain man to owner of great stretches of ranch land on the Rio Grande, populated with several hundred head of horses and cattle. Within a year of their marriage, Augusta, his German bride, began bringing into the world, in quick succession, a son and three daughters. This she did in primitive conditions and without a doctor. Augusta's role was essential—she was the person who brought to this wilderness of sagebrush and cattle a home, a family, and a reason to build up. As his family grew, Fritz Becker added on more acres of land, enlarged his cattle herd, and constructed a school on his land for his and the neighbor's children. In time, he added a cemetery where he could bury his own, if necessary, and maybe even a stray hired man or two.

Peter Hansen laid down a similar foundation for his son and two daughters—building his eventual fortune in much the same way Fritz Becker had. Having immigrated from Schleswig-Holstein, Germany, as a boy of nineteen in 1856, Peter found his way to Colorado about the time of the outbreak of the Civil War. He enlisted in Company 1-First Colorado Infantry and went with his regiment to turn back the Confederate Army's crusade into New Mexico at Glorieta Pass. By the end of the war, he had a somewhat more innocuous job—that of tending Fort Garland's dairy herd on the Conejos River in the southernmost part of the San Luis Valley. After his discharge from the army, he stayed on the Conejos and began what was to be a lifelong process of buying and building up ranches. It was rough going, but he succeeded at it and eventually passed on to his son and two daughters a fortune in land and cattle—one that he had started with fourteen milk cows and a crude adobe hut on a little river on the edge of nowhere.

Peter Hansen spent a decade on the Conejos, where he had brought his bride and where his son and first daughter were born in the early 1870s. He eventually relocated at the confluence of the Conejos and Rio Grande Rivers and, by the time his second daughter was born, he had established what would always be called the "River Ranch." Emma and Lizzie Hansen's first home at the River Ranch was a modest three-room log and sod house. The ranch itself was situated along the west side of the Rio Grande River, a lovely setting shaded by a large grove of cottonwoods—a picturesque place of corrals, bunkhouses, and big barns with great hay lofts. Later a large, prosperous looking seven-room ranch house fronted by a big bay window was built. Flower beds and shrubs enhanced the large yard. It was a house of substance and said to be one of the finest ranch houses in the valley. The ranch's artesian wells and plentiful river water were enlivened by the constant presence of water fowl—ducks and geese. The Hansen's River Ranch was an isolated paradise—a place where the sky burned with red sunsets at evening and the hills glowed with wildflowers in the summer. The finest horses and the best cattle—Galloway and Hereford—grazed the pastures throughout the year.

"Papa Pete" Hansen, being very much a "man's man," expected a full measure of masculine accomplishment from his only son, but possibly spoiled and pampered his two daughters, Lizzie and Emma. By the time they were teens, their prosperous father had handed them an enviable package—independence, wealth, beautiful surroundings, and the rich heritage that went along with ranching. All this was theirs for the taking. It would have been surprising if the girls, seemingly born under a lucky star, ever wanted to leave such a charmed life. Their mother, in fact, had predicted that there were not likely men around who could ever give her girls what they had at home. Her suspicions proved to be true—neither Emma nor Lizzie Hansen ever married. They spent their lives in the occupation of ranch women on Papa's big spread. Unfortunately, they both died at a relatively young age, but it couldn't be said that they didn't live their lives to the fullest.

Many of the Hansen girls' early pictures show them riding horses. They were somewhat isolated on the ranch and relied on each other and their horses for companionship—as many a ranch girl has. Horses became a lifelong preoccupation for Lizzie, the more "outdoor" of the two girls. Of course, along with horseback riding and participation with other ranch activities, the girls had to go to school. Their early education was completed at the River Ranch with private teachers. Later, their wealthy father sent them to a finishing school in Chicago. Ever more polished afterwards, they went home to the ranch and got back on their horses. A taste for broader

Chapter Seven The Becker and Hansen Cousins of San Luis 91

Lizzie and Emma Hansen riding sidesaddle at the River Ranch around the turn of the century.

Photo courtesy of Eileen Hansen Burt.

culture must have stayed with them, however, for in later years, both girls traveled the world, seeing Europe, Canada, Cuba, Mexico, Puerto Rico, and China. But no matter how far they journeyed, they always returned to the River Ranch. At Papa Pete's death in 1913, his property of over 7,000 acres was divided among his family. His will provided for Emma and Lizzie to each receive a substantial amount of land. After his death, they took charge of the actual operation of their inheritance. Some reports credit Pete's son, William, with overseeing both his sisters' and his mother's property and livestock after the death of the family patriarch. There can be no doubt, though, that Emma and Lizzie, for several years, had a major part in the management of their property—Lizzie taking care of much of the outside work and Emma keeping the ranch's books.

The Hansen girls' halcyon days astride horses on Papa's big ranch were glorious but all too short, lasting a brief twenty years of adulthood before ill health plagued them both. Lizzie, born in 1873, suffered from pernicious anemia for seven years and died at the age of forty-nine. Emma, born in

1875, succumbed to breast cancer at the age of forty-seven after a six-year illness. She died within months of her sister. The vigor and long life of their pioneer parents were not to be theirs. How hard it must have been for Emma and Lizzie to say good-bye to their charmed life too soon—to pine away, as Lizzie did, in a Denver hospital, longing for the sight of a hawk circling down the breeze over the San Luis plains, or the smell of the sweet clover drifting in from the meadow.

While Emma and Lizzie Hansen had been enjoying their good ranch lifestyle and world travel, their nearby cousins—Anna, Emma, and Lizzy Becker—were living a similar but more run-of-the-mill existence at the Becker Ranch. They, too, had been schooled at home in a one-room schoolhouse, while at the same time learning the trade of ranching. Emma, Anna, and Lizzy Becker all took a very active part in the operation of their father's ranch before marriages altered their careers. Haying, riding the range, branding, irrigating, and calving were jobs they had to tackle. When the girls' brother, seeking independence, left home to work for wages at ranches

Lizzie Hansen in later years, riding astride and working cattle at the River Ranch.

Photo courtesy of Eileen Hansen Burt.

some distance away, the girls pitched in and took his place. They worked right alongside their father and were his major source of labor. They weren't as "hard pressed and put upon" as many might have thought; they relished being cowgirls and got pretty good at it. But it didn't last long—the girls soon began leaving the ranch one by one to get married.

Emma Becker, the youngest, born in 1878 on the Becker Ranch, married a prominent local cattleman, Fritz Emperious, and had two children. After her marriage, she continued to sometimes "tend to things"

Lizzie and Emma Hansen—proud and dressed in high fashion—but still ranch women.

Photo courtesy of Eileen Hansen Burt.

at her father's ranch, seeming to find these obligations hard to leave behind. She lived out her life in the San Luis Valley, and outlived her husband by twenty years. When she died in 1956 at the age of seventy-eight, she was honored at her funeral by an old Western tune befitting a cowgirl, a rendition of "Springtime In the Rockies." Anna Becker, born in 1872, also married a rancher—William F. Pim—had a son, and outlived her husband by several years. After residing for a few years in nearby Creede, where they were engaged in the mining business, Anna and William returned to a ranch near Alamosa. Anna went back to work riding the range, as well as doing every other task required on the ranch. She died at the age of eighty-one, having spent seventy-five of those years in the San Luis Valley. Born in 1873, Lizzy Becker, also known as "Eliese," married Charles Meyer and resided near Alamosa. She had two children who were still quite young at the time of her unexpected death at the age of forty-four in 1917. Until she died, she was actively involved in running the family ranch. During their lives, the three Becker sisters were able to accomplish something that their Hansen cousins had been unable to: they had produced children, sons and daughters who would later tell of their mothers' rich experiences as some of Colorado's very first authentic ranch women.

For the most part, the Becker and Hansen girls never ventured very far or for very long from the San Luis Valley and the ranches they had been born to. They had written a new chapter in the West's history—one for women that was unprecedented—and one that should not be forgotten. But the cousins with the same names live on mainly in family reminiscences; otherwise, they are not well-known. And the Becker and Hansen ranches, so much a part of their story, are no longer ranches. The barns, ranch houses, and blacksmith shops are long gone; the sagebrush flats and the hay meadows are all now a part of the Alamosa Wildlife Refuge. The land of the Becker and Hansen ranches seems to have gone back to its origins. Now there is hardly a trace of the five women who once wore the title of "ranch woman" with pride, rode bulls in sidesaddles, and branded in their "Sunday best."

Left to right, Anna Becker Pim, a relative,—Betty Emperious, and Emma Becker Emperious in later years.

Photo courtesy of Eileen Hansen Burt.

Chapter Eight
ELIZABETH

Elizabeth Bassett
(1855-1892)
Bassett Ranch — Brown's Park, Colorado

> *"Her own outfit consisted of a beautifully fitted 'habit' of rich, dark blue material, long skirted and draped with grace. For trimming there was a number of gleaming brass buttons. She was a blond, five feet, six-and-a-half-inches tall. Mounted on her thoroughbred saddle horse, 'Calky,' she was a picture to remember."*
>
> Anne Bassett Willis of her mother, Elizabeth Bassett
> *Colorado Magazine*, April 1952

Brown's Park in the late 1800s had not seen too many white women—had never seen one like Elizabeth Bassett—and probably never would again. Northwest Colorado's wild and isolated Brown's Park and the gentle Southern belle Elizabeth Bassett were quite an unlikely match—a match that one would not have expected to last, much less flourish—but flourish it did. In time, Elizabeth's two daughters, Josie and Anne, would become more famous (and in some opinions, infamous) than their mother would ever be. Elizabeth Bassett's daughters, as grown women, would manage ranches, ride and shoot like regular cowboys, and—according to some accounts—rustle cattle and romance outlaws as well.

Elizabeth's two daughters did, indeed, gain more than a little notoriety in their lifetimes. As a result, much has been written about them. Their mother, Elizabeth, however, was the one who paved the way for them—laid

Chapter Eight Elizabeth Bassett of Brown's Park

the groundwork that made possible the rich experiences her girls would have later on. Notoriety was no part of it for Elizabeth. One of her faithful and devoted hired hands—a well-known cowboy of the old West, Isom Dart—said that Elizabeth Bassett possessed a "Southern quality." Men such as Isom Dart tended to appreciate women who were beautiful and well-bred to boot . . . but possibly underestimated them. Elizabeth Bassett, for all her gentle upbringing, blond beauty, and Southern charm, was tough as nails.

Brown's Park, where the Bassett family chose to settle, could be described as tough as nails, too. Located in the northwest corner of Colorado and encompassing some of Wyoming to the north and Utah to the west, it was a place of pure mountain beauty. Early on, the park had been home to various Indian tribes and, later, was frequented by the mountain men and fur traders for their rendezvous. Later still, in the 1830s, it was the short-lived home of Fort Davy Crockett—often referred to as "Fort Misery." It was also a longtime location for wintering herds of cattle. By the time the Bassetts arrived, the park was well on the way to becoming "cattle country"—not just a winter holdover for herds headed north and west, but a permanent home for cattlemen. Cattle outfits, whose sizes ranged from the huge acreages of cattle barons down to the quarter sections of sod busters, would soon inhabit Brown's Park. The park also experienced, with the rest of the young West, its share—maybe more than its share—of cattle rustlers, horse thieves, and outlaws. "Necktie parties," as hangings were sometimes called, provided quick justice for those with the long ropes and the running irons. The road through the park was known to be part of the "Outlaw Trail," and had seen the likes of Butch Cassidy and his Wild Bunch. Eagle-eyed scout and Indian fighter Kit Carson had once inhabited the park, as had beady-eyed Tom Horn, pseudo cattle detective for the Wyoming Stockgrowers Association, who was hung in that state for the ambush and killing of a fourteen-year-old boy. But Brown's Park in the late 1800s, having seen just about everything else, had never seen anything remotely resembling Elizabeth Bassett.

Most of what is known of Elizabeth's life has been gleaned from several articles written by her daughter, Anne Bassett Willis, and published in the 1950s in *Colorado Magazine*. Mary Eliza Chamberlin Miller, Anne's mother—always known as "Elizabeth"— was a daughter of the South, born in the mid-1800s. Her parents died when she was a child, and she and her sister were sent to live with their grandparents, who raised them to young adulthood in Norfolk, Virginia. They grew up accustomed to the comfort bestowed on girls of "good family." Even as a young child, Elizabeth was known to be gracious and ladylike as would have been expected, but she was also possessed of quite a strong will—an iron will, according to some—which at times could be channeled into a very hot temper. Elizabeth was married at the tender age of sixteen to a man twenty years her senior. Her choice of a husband was a man who might well have been considered "fatherly." But the age difference proved to be no hindrance to the marriage, which from the beginning was a strong union—a happy and lasting one. When he married Elizabeth, Herb Bassett was a Civil War veteran who had held the rank of major. He was described as being handsome, well-educated, musically talented, and very religious. Possessing a somewhat passive personality, he was a gentle, kind, and fatherly man who would always be held in high regard by his young wife.

Herb and Elizabeth, shortly after their marriage, which, according to Anne, took place in September of 1868, moved from Virginia to Little Rock, Arkansas. Here, Herb held a job as Clerk of the Court, and Elizabeth bore two children—Josie in 1874, followed soon by Samuel. Their idyllic Southern life might have continued on in Little Rock had it not been for Herb's poor health, thought to have been brought on by either malaria or asthma, or both. The couple was advised that a change of climate would improve Herb's health, and that advice was responsible for a decision to move to the semiarid West. By the late 1870s, the family had migrated by train to Wyoming, where they stayed for only a short period of time. They then moved on by wagon to Brown's Park to join forces with Samuel Bassett, Herb's brother, who was already well-established there.

Leaving her life of comfort behind, Elizabeth Bassett arrived to an uncertain future in Brown's Park with an ailing husband, two young children in tow, a third one on the way, and most of her worldly possessions piled in a wagon. Uncle Sam Bassett's small dirt roof log cabin served as home for the whole family that first year. A band of Yampatika Utes camping close by were their nearest neighbors. There were also a few other settlers spread out over the park, and this odd assortment of people made up Brown's Park's early population.

Uncle Sam Bassett had been among the first of these settlers and was, therefore, one of the most familiar with the park. As a young man, he had seen much of the West as a prospector and, eventually, a government scout along the Overland Trail. In 1852, he spent some time in Brown's Park in the company of mainly Ute Indians, and he eventually decided that the park was a good location for a permanent residence. In Uncle Sam's rough log cabin, Elizabeth gave birth to the first white child born in the park, bringing daughter Anne into a world that could only be described as primitive. Many other pioneer women did the same—some even giving birth to children in covered wagons. Not many did as Elizabeth did, though. She found a Ute woman to serve as a wet nurse for Anne when her own milk proved to be inadequate.

Eventually, the Bassett family moved from the hospitality of Uncle Sam Bassett's cabin and found a place of their own in the park, a prime location determined to be a good spot to start building up a ranch. The ranch house or cabin was described by Anne as being a log house built around 1878, a low rambling structure that stood near a spring and at the foot of Cold Spring Mountain. From this location, one could see picturesque Ladore Canyon, the entrance to the park at the east end of the valley. In the shelter of close hills and cottonwood trees, the family pitched camp while the five-room cabin in the shape of a T—or cross—was being built. A garden was planted and a few cattle purchased. The result of the family's efforts was the usual Western ranch's rough and primitive beginnings. Over the years, the place blossomed with the addition of barns, bunkhouses, sheds, corrals, and—in time—even an orchard. The Bassett ranch was on the main road into the park and became a stopping place for many of those first entering it. And, according to the West's unwritten code of hospitality, these unannounced guests were always made welcome and usually fed. The Bassett's log cabin was furnished with graceful old family furniture brought from the South, as well as an impressive library. Mixed in with the formal furniture were bear and buffalo rugs and furniture made by Herb out of rawhide and birch. A huge cookstove, a small portable organ, and always the presence of small lively children made for an atmosphere that was charming and comfortable. It was a genuine ranch house, a real Western home—one with many windows and several warming fireplaces.

None of this—not the trip west, the homesteading venture, or the friendly Indian neighbors—was out of the ordinary or much different than that which was experienced by the hundreds of other families settling the West. What was decidedly unusual about the Bassett's situation was

Bassett cabin belonging to Josie when this photo was taken.
Photo courtesy of the Museum of Northwest Colorado.

Elizabeth's fascination with ranching and her willingness, in spite of the burdens of home and children (two more would come along), to jump head-first into the totally foreign business of ranching—very unknown to her and very much a man's domain. Her husband, while being supportive of this venture in many ways, from building furniture for the house to planting an orchard, was said to have never really been much of a rancher, nor did he want to be. His interests were elsewhere, with the affairs of town and neighbors. In time, he became heavily involved with community efforts of various kinds. He even took it upon himself to conduct church services from time to time, as the community was often without a minister. It was Elizabeth who took to ranching like a fish to water. With that rare combination of traits—womanly softness and a backbone of steel—Elizabeth seized the opportunities she saw before her and began breaking down barriers and building barbed-wire fences.

Elizabeth quickly learned how to use a rifle and became skilled enough with it to shoot game. She already knew how to sit a horse, even though her ladylike sidesaddle riding must have been quite a hindrance. What she didn't know about ranching, which was just about everything, she eagerly learned and was soon moving range cattle to better grass, doctoring sick cows, and packing salt out on the range like a veteran cowboy. Wisely, she had no objection to hiring capable help, and in time her bunkhouse became home to many a good cowboy. Whatever ranch operation was going on, however, Elizabeth always joined right in with the work, no matter how difficult. She was the undisputable boss, but much of her instruction was accompanied with a charming sweet smile. The hired hands seemed more than eager to please her. It was often said that her cowboys would have willingly died and gone to hell for her; it wouldn't be a far stretch to imagine that a few of them were even half in love with her. And Elizabeth treated these ranch hands just like family, knowing how valuable they were. What a picture it must have been—rough-and-ready cowboys playing cards in the bunkhouse, roughhousing with the Bassett children, doing household chores such as building fires and carrying water—even cooking.

When Elizabeth arrived in Brown's Park, inexperienced though she may have been, she immediately saw the great potential of the cattle business there—saw that she could "turn it to her advantage," as her daughter put it. The park, some thirty miles long and six miles wide, had first been a holdover for the beef cattle being trailed to the west coast. The winters were unusually mild there, with only light snowfall. The range cattle business in the park, up until the Homestead Act of 1862, was pretty much run on the concept of free grass where "the firstest got the mostest"—a case of unwritten laws and unfenced range. This practice allowed the large and prosperous cattle outfits to become huge and sometimes threatening, taking full advantage of all the free grass they could find. The owners of the large cattle outfits, sometimes referred to as "cattle barons"—while able to help themselves to much grass, also had their share of problems. They fought off cattle rustlers and the hated migratory sheep outfits that came into the park in large numbers. These huge moving bands of sheep had the capability of destroying entire ranges in their wake. Where the law was inadequate or had not yet arrived, the cattle barons took it into their own hands—wiping out great numbers of sheep with bullets and clubs and hanging cattle rustlers on the branches of cottonwood trees or the roof beams of barns.

Into this picture came homesteaders such as the Bassetts, who—in accordance with the Homestead Acts—staked legitimate claims on land heretofore up for grabs. They weren't exactly welcomed, and many and long were the battles that resulted. Time and again, the big outfits and homesteaders such as the Bassetts did not see eye to eye over what range belonged to who. The mountains and mesas, the sagebrush hills and grasslands that made up Brown's Park became hotbeds of dispute over personalities, territories, and philosophies. In time, the Bassett family would take part in more than one such war.

What a challenge this must have been for a woman like Elizabeth to take on—this "code of the West cattle raising"—and all the while trying to raise several children. Even with the help of the eldest girl, Josie, family responsibilities alone must have been a tall order in such primitive circumstances as existed in the park in the latter part of the nineteenth century. But not for a minute would Elizabeth Bassett give up on the excitement of creating her own little empire of water and grass and cattle—even in the presence of strong competition. She began building up a ranch, adding on land and building up a cattle herd—supervising and assisting with much of the work. Her daughter, Anne, described her in these years as being a small woman physically, but one who didn't know the name of fear—gregarious, out-going, outspoken, heartily liked by those who liked her, and one who claimed an almost fanatical loyalty among the other Brown's Park settlers.

And so it went for Elizabeth Bassett—this challenging and unconventional life—but for too short a period of time. In 1892, when she was in her mid-thirties and at the height of her ambition and success as a

Elizabeth's daughter, Anne, in a turn-of-the-century era dress.

Photo courtesy of the Museum of Northwest Colorado.

rancher—at a critical time in the lives of her five young children—suddenly and unexpectedly it all came to an abrupt end. Elizabeth was struck down with an illness that one daughter remembered as being an appendicitis attack, and another daughter suspected was a miscarriage. Whichever it was, in a short two-week time period, Elizabeth was dead of a cause that a little medical attention may well have prevented. Even though her death was an irreplaceable loss that devastated the family, they found it was necessary to go on with the business of running the ranch. This was accomplished with the assistance of hired men, while a hired housekeeper took on the impossible task of replacing Elizabeth as the woman of the house. One of Colorado's most authentic ranch women, as well as one of its first, was buried on a hillside overlooking the ranch she had all but created. Here, overlooking the silver-green sagebrush, the summer hills radiant with wildflowers, and the winter prairie dusted with snow, Elizabeth lay alone and kept silent guard on her ranch and her family below.

Ranch woman Josie Bassett, Elizabeth's oldest daughter, in her senior years.
Photo courtesy of the Museum of Northwest Colorado.

One wonders if life might have been different for the remaining Bassett family had Elizabeth lived. Would her two high-spirited daughters have been a little less controversial had her ideas of genteel womanhood held influence over their teenage years? Would Herb Basset have died of old age with his wife by his side instead of alone in an old soldier's home far from Brown's Park? Would the Bassett ranch have thrived more under her influence? Would she have received the recognition she so well deserved for leading the way for women—for showing them that the West was a land of opportunity, even if one had no experience to bring to it and wasn't a man?

Anne Bassett around 1912 in Sonora, Mexico.
Photo courtesy of the Museum of Northwest Colorado.

Chapter Eight Elizabeth Bassett of Brown's Park

No photographs seem to exist of this fair daughter of Brown's Park, but Anne and Josie Bassett, both quite pretty women, must have resembled their mother. Anne Bassett Willis, in her memoirs, spoke eloquently of Elizabeth, and her words must serve in the place of a photograph: "Mother was a woman of truly distinctive personality, with many remarkable qualities. From childhood she had been required to do nothing more fatiguing than to summon a negro slave to perform even the slightest tasks for her. But she neither faltered or gazed longingly back to those early experiences when her life's connections were broken by the Civil War. She looked ahead, seeing adventure and alluring excitement as my father's helpmate and companion in the new West."

Chapter Nine

BARBARA

Barbara East
Western Slope Range Rider and Artist
Boulder, Colorado

> "Somewhere within the vast miles of cows
> and country... my dream of being a
> range rider was born."

Barbara East
Gunnison County Stockgrower's 100 Year Anniversary Issue, 1994

Barbara East is a pretty woman—maybe not the type seen on the covers of fashion magazines, airbrushed and with "big hair" and perfect makeup, and maybe not even the contrived type seen in Western magazines who glitter with turquoise and silver and wear buckskin with fringes, portraying a Santa Fe look. It takes about five minutes after seeing Barbara out on the range on her horse for your mind to register the fact that she is a pretty woman. The first impression one gets is that she is rather dusty, but the dust goes along with the natural grime and horse sweat that sticks after a ten-hour day of hard riding. Her battered and well-worn hat pulled low on her brow hides her face. When she takes off the hat and starts to talk, that's when the beauty comes into view—the natural beauty of a strong face toasted a little brown by the sun, lined just a little by life, and lit up by pure excitement about cows and horses and mules and mountains—riding the range.

Barbara East is a resident of Boulder, Colorado, but that's not where you'll find her in the summer months. She'll be on any one of a number of mountains on the Western Slope—herding cattle. She has spent over thirty summers off somewhere in the Rocky Mountains doing just that. It's what she loves best. Winter will find her back home in Boulder at an easel with a paintbrush in her hand, creating on canvas what her eyes have taken in all summer. Barbara is a professional range rider and a first-class Western artist—quite a rare combination of talents—and in both of these endeavors she has achieved remarkable success.

One might expect to find an artist in Boulder, but it seems an unlikely place to find a cowgirl. This progressive Eastern Slope city is known more for its college coeds and idealistic environmentalists. Yet, Barbara East is hardly out of place here. She grew up in Boulder with neighbors who were ranchers and farmers, a small minority of folks who held together the agricultural segment of this foothills town that was once little different from any other Colorado frontier village. Here in the mid-to-late s1800s, former Indian lands began to be occupied by miners and, later homesteaders who had ranching and farming in mind. The new inhabitants struggled mightily to make a success of agriculture, growing alfalfa, fruit trees, and vegetable gardens, and raising cattle who thrived on the valley's great grasslands.

The beautiful Boulder valley is know for its mountain backdrop—the leaning Flatirons out west of town. The pristine beauty of this area's canyons, hills, mountains, and streams seemed somehow to cry out, even early on, for preservation by those who resided here. Even at the turn of the century, its residents had their eye on much of the early-day landowner's holdings and government land, as well. Where old homesteads once existed are now found thousands of acres of preserved open space, which are shared by recreational interests and a few ranch families who lease some of this land for the grazing of cattle and the production of hay. And so, a small contingent of ranchers is still to be found in Boulder—many of them descendants of the original ranch families. Among them is Barbara East, who relishes her "rough around the edges" friends of the "Old West," while holding her own among those "more cultured" of the "New West."

Barbara East has a summer job that hardly any other woman could comprehend. Her office is a truck cab, her work space an intimidating 50,000 acres, her work load seven days a week, ten hours a day. Her primary job description is the moving, salting, and doctoring of cattle. Her secondary job description includes pleasing the public, ranchers, and environmentalists—often at odds with each other—who use the public lands that make up her cattle range. Barbara's co-workers are cow dogs, mules, and horses. Her work clothes are jeans, slickers, boots, chaps, and spurs. Out on the range, she doesn't have to worry about phone calls, though—cell phone signals are usually nonexistent where she is working. Sometimes she works the night shift, too—that is, if moving cattle by moonlight is easier on the livestock than moving them in the heat of the day. Hardly anyone in the modern world—other than a few hard-edged cowboys—would understand Barbara East's job.

For the past two summers, Barbara's range riding has taken place on Grand Mesa, the beautiful mountain jutting into the sky east of the city of Grand Junction. It's the largest flattop mountain in the world and has traditionally been used for livestock grazing by both sheepmen and cattlemen who have shared its rich treasures of grass during the summer months. Barbara's territory is pretty much within the confines of the Grand Mesa National Forest, where local ranchers have permits to graze cattle. This rugged county is inhabited by an interesting mix of cattle, tourists, fishermen, hunters, elk, and deer, along with an occasional black bear, moose, or mountain lion. Here is where Barbara keeps watch over the cattle in her care, always keeping a close eye on them; she's bent on being fully apprised of their whereabouts, always conscious of her responsibility to the permit holder (rancher) who owns them. On any given day, Barbara might move these cattle, deliver them salt blocks, repair fences, clean a trail, appraise the health of the range or the availability of water, and rope a sick calf or two and give them shots of an antibiotic to stop a case of pneumonia. On top of all these chores, she begins and ends each day caring for her own animals—her numerous horses, dogs, and mules. Barbara has a house located some ten miles downhill from the camp on the top of Grand Mesa. She leaves home early each morning and often returns long after dark. Back home in the Plateau Valley, she finds the welcome security of a telephone and checks in with a friend, just to let someone know that she has returned home safely another day.

Barbara brings a lot of experience to her Grand Mesa job, but—even so—she claims that it has taken her two years to learn what, for her, is new country. In her care are 900 to 1,100 cow and calf pairs, cattle belonging to half-a-dozen ranches located nearby. Every year this livestock is moved from home ranches to higher ground, pooled together in one group, and supervised by a range rider such as Barbara. She has a definite operating plan in mind for the cattle in her care on this mountain; nothing she does is haphazard. One of her goals is to maximize cost-effectiveness for her employers—one more way to promote the long-term stability of the cattle operation for those ranchers who put their livestock in her care. She has the practice of managing such vast country and huge numbers of cattle down to a science. Taking into consideration the differing philosophies that sometimes exist between the owners of the cattle and the personnel of the United States Forest Service, Barbara is able to perform the small miracle of juggling and attending to it all, making everyone mostly happy at the end of the day. She likes the diversity of her group of employers—ranchers who range in age from twenty to ninety and have an interesting mix of traditions, plans, and ideas about cattle ranching.

Grand Mesa is only one of the mountain and cattle ranges that Barbara has had to master in her thirty-four-year career as a range rider. Most of that time has been spent in the Crested Butte area where, for twenty-six summers, she alone was responsible for the care of up to 1,800 head of cattle spread over a huge region that included, at one time or another, the grazing areas associated with Slate River, Oh be Joyful Creek, Brush Creek, Pearl Pass, Gothic, West Brush Creek, and Deer Creek—in other words, most of the country surrounding this famous little Western Slope ski town. In Crested Butte, Barbara sometimes herded cattle through fields of wildflowers growing under the chairlifts that were empty in the summer months. Once, this village of condos, ski lifts, and trendy boutiques was simply a mining town surrounded by ranches and cattle ranges. A century ago a group of early-day cattlemen, calling themselves the "East Side Cattlemen's Association," organized, pooled their cattle, and extended their reach into all the high ranges of the area. Here, they ran up to 5,000 head, occasionally moving a few of them down Elk Avenue (main street) and down Highway 135 to home ranches.

Now, more than a century later, Crested Butte is hardly cattle country anymore, and at the end of her twenty-six years of riding the range there, Barbara found herself increasingly competing for use of the range with recreationists and others, some bent on seeing cattle removed entirely from public lands. And so, after a long love affair with this country, she moved on to try her hand in other locations. Other western Colorado ranges that have been workplaces for Barbara are the Cebolla range, south of Gunnison; Saguache Park, southeast of Gunnison; the Cimarron range, southeast of Montrose; and Muddy Gap, located between Walden, Kremmling, and Steamboat Springs. It was on the Cimarron range that Barbara

Barbara East at the Trampe Ranch in Gunnison, Colorado, October 2005.
Photographer, Walt Barron.

experienced what she called her "wildest summer," trying to keep track of 1,500 wild Black Limousin steers in eight miles of solid oak brush. She claimed they were crazy—jumping corrals and breaking legs. They could hardly be handled by Barbara (or anyone else, probably) without heavy use of "dog power"—the canine "professionals" trained specifically for the tricky and dangerous job of keeping such unrestrainable animals under control. These dogs are naturally inclined to outsmart and outwit cattle—engaging in a war of wills between cow and dog that they and Barbara usually won.

Although Barbara has worked cattle with the assistance of others—usually men—she prefers to do the job alone. A strong advocate of the "low stress" method of handling livestock, she abhors the rough handling of cattle or horses that has all too often been a tradition in the West. Barbara claims there are three words to remember when handling livestock: "patience, patience, patience." Working alone and working slowly and calmly is her method; trial and error has been her teacher. Along with mastering the art of handling livestock in this manner, she has also, over the years, honed her skills in land management. At this point, whoever hires her can be assured that their grazing permit on the national forest will be well taken care of, as well as their livestock. Ranges damaged by overgrazing are not tolerated by Barbara. By now, her reputation as a range rider is sterling, and seasoned old ranchers—who, not too many years ago, would have laughed at the prospect of a woman range rider—jump at the chance to put her on the payroll.

However rewarding this job is for Barbara, when summer wanes into fall, the last of the aspen leaves have littered the ground with gold, and the cattle have been returned to their home pastures, it's time for her to go back to the "civilized" world of the Eastern Slope and Boulder. It's undoubtably an adjustment to leave the beautiful mountain trails and the slow moving of lazy cattle behind to face the snarling traffic back on the Front Range. It's hard to say which place seems more like home for Barbara but back in Boulder, she returns to the old East family house and begins the process of adjusting to a more confined style of life. It starts to fall into place when she brings out the easel and paintbrushes—tools of her second occupation. And it is not technically "city life" that she has come home to, anyway. Barbara has an agricultural history in Boulder and represents the fourth generation of her family there. Her grandparents started the tradition, her grandfather being a fire chief who housed the fire horses in a stable next to his house in the early days. Her grandmother loved riding horses, and her father was a "gentleman farmer" who was further tied to the agriculture of the area through the ownership of King and East Machinery, a farm implement business in Denver. While Barbara didn't grow up on a ranch, she was nevertheless well acquainted with saddle horses and fire horses as a girl. She says she experienced the "spirit" of ranching without being directly involved in it. She claims to take after her mother in many ways—a woman who was a housewife, and one who had a great sense of humor and pride, staying power, and an appreciation for a sense of accomplishment. Barbara interrupted her career as an artist for the last several years to care for her mother as she slowly declined from Alzheimer's disease. It is in her mother's house that Barbara lives now in the winter, doing all the outside work required on the few acres of the family home while staying in touch with livestock by feeding the neighbor's cattle. Barbara also owns several head of horses and mules and sixty head of her own cattle, a type known as Black Angus crosses. She keeps them in Gunnison and returns to the Western Slope often to care for them.

Barbara's art career began in the early '70s, while she was attending the University of Montana and working under the guidance of famous freelance artist, Ace Powell. At that time, Ace was getting up in years, and he enjoyed helping aspiring artists get their start. As a young man, he had learned the trade by working with Charlie Russell, the famous Western artist. In time, Ace Powell became well known in his own right for his oil paintings of the cowboy genre and his Indian figures, all done in his vivid and realistic style. He also sculpted in stone and wood, and his trademark ace of diamonds signature finished all his work. Barbara's artistic skills were enhanced by this talented mentor, helping her get her start in the marketplace.

Barbara East with her cow dogs at the Trampe Ranch in Gunnison, Colorado, October 2005.
Photographer, Walt Barron.

Over the years, Barbara's art has continuously improved and her work has sold well. She goes about producing a painting with the same work ethic used in riding the summer ranges. She feels that how well one succeeds in any endeavor is relevant to how hard one works. Barbara's art work, having sold all over the West, Mexico, and Canada, is a testament to that hard work. She sells on private commission, commanding as much as $3,000 for her miniature watercolors on cigarette papers—an idea she got from her old friend and teacher, Ace Powell. Her larger works are extraordinarily good—exquisitely detailed scenes of horses, harness, cattle, cow dogs, and mountain scenery. Barbara can capture the exact expression on a

Chapter Nine Barbara East of Boulder

This painting by Barbara East is a self-portrait depicting the difficulty of moving cattle in rugged country. Courtesy of Barbara East.

weary cowpuncher's face or the wary anticipation of a cow dog crouched and ready to go to work herding cattle. Newer subject matters are evolving now for Barbara as she tries her hand at detailing the smaller things of nature. Gourds, grass, pine cones, and leaves are exacted by her in egg tempera, a very old and difficult medium.

Never satisfied with her considerable achievements, Barbara continues taking professional classes in the winter to upgrade her artistic skills and claims, "You can always get better. If you quit looking to get better, then you're as good as you're ever going to be." At the same time she stays abreast of the latest developments in public lands grazing, always looking for ways to improve things next summer back on a mountain somewhere. Never one to be confined to a small world, either as an artist or a range rider, Barbara has plans to someday try out more new country, perhaps in Wyoming, hiring on where she can find an untried range to ride. While Barbara returns home to Boulder each winter, she also spends time in Gunnison with a rancher who has been her long-time friend and companion. Barbara seems to find life to her satisfaction all the way around.

Barbara East is anything but a traditional woman. She knows her two occupations do not lend themselves to family life—that one cannot "have it all." Yet, she is too busy to get lonely. She knows that her living conditions in a cow camp are primitive, at best, and that her pay for the range riding job isn't up to par with the experience and skill it requires. Of her range riding, Barbara said in the February, 2003 issue of *Rangelands* magazine, "I think of it as a discipline and the study of patience in a clockless workplace, where the pace of a cow and the track of the sun measures the rate of time." Most women—and men, for that matter—wouldn't want her job . . . couldn't do her job . . . and yet almost everyone envies her.

Chapter Ten

VEVARELLE

Vevarelle Outcalt Esty
(1902-1999)
Esty Ranch — Gunnison, Colorado

> *"You don't find true happiness in life until you find it in your work."*
>
> Vevarelle Esty, 1993

In order to understand Vevarelle Outcalt Esty, one would first need to envision the father she so dearly loved. Gunnison County's pioneer rancher, John Outcalt, made a lasting impression on his youngest daughter and served as her guiding light all through the nine decades of her long life. John Outcalt apparently had quite a gift with children—earlier with his younger brothers and sisters, who had been left in his twelve-year-old care upon the death of their mother, and then later, with his own.

Vevarelle liked to reminisce about her father—recalling his arrival from Brunswick, New Jersey, to the city of Denver in the early 1870s at the age of twenty-one, armed with the skills of a ship's apprentice, a tool box, and twenty-seven cents in his pocket. According to Vevarelle, her father, shortly after his arrival, acquired work as a carpenter in Denver. While participating in the building of that early Colorado city, he made the acquaintance of a man who had great interest in what was known as the "Gunnison Country" on Colorado's west slope. Sylvester Richardson—variously known as a geologist, druggist, assayer, teacher, lawyer, and historian, among other things—upon first seeing the undeveloped Gunnison valley, had became greatly excited about its agricultural possibilities. John Outcalt and his brother became part of a "colony" organized by Richardson in 1874, and made the trek over the mountains with the rest of the group to look over the raw and uncivilized new country on the other side. The flamboyant and eccentric Sylvester Richardson dreamed big of turning the Gunnison valley into an agricultural Mecca—utilizing its vast reaches of grass, which were ideal for the grazing of livestock. The Gunnison

Country was located in the center of the Rocky Mountains at an elevation of 7,703 feet and was well known for its severe winter weather. In a typical winter, deep snows and extremely cold winter temperatures lengthened into a nearly nonexistent spring. Isolation, high elevation, long winters, short summers, and distance from markets were all shortcomings of the country's agricultural prospects. But on the other hand, there was such abundant grass and plentiful water that Richardson and the others were convinced that with a little stubbornness and determination on their part, this beautiful mountain-locked country could be tamed.

Pioneer rancher John Outcalt in his early days in Gunnison, Colorado.

Photo courtesy of the Esty family.

Upon their arrival, the Outcalts found the country, indeed, very cold for Richardson's proposed ranching or farming, and they found no other white men in the area except the occupants of a government cow camp on the Gunnison River—one that supported the Los Pinos Indian Agency that was located nearby. But they were heartened by the country's prospects, as well as its great beauty. Undaunted by the hardship that was surely ahead, they proceeded to secure property located five miles northeast of the present-day site of Gunnison on the Gunnison River. The two brothers were young and vigorous, and the pioneer spirit burned strong and it wasn't long before they were harvesting 500 tons of hay and had over one thousand acres of meadow land under production. It was wild hay, but they considered it as good as tame hay, stating that it only needed irrigation, then "the hay would grow itself." Then they turned their attention to cattle raising, trailing one of the

The first cabin built on the Outcalt homestead.

Photo courtesy of the Esty family.

first private herds into the Gunnison Country over a mountain pass. While John Outcalt's brother tended to the ranch, John added to the ranch's coffers by spending two years in the employment of the government at the Los Pinos Indian Agency, located southeast of Gunnison near Cochetopa Pass. In the early 1880s, when the government job ended, the Outcalt brothers added to their ranching enterprise by establishing a business in the newly incorporated town of Gunnison. They built a livery stable, which they called "Free Corral." Here, they sold mowers, reapers, and binders for a "live and let live" price, as well as snake oil remedies and even baths, all reasonably priced. The livery stable thrived and became so prosperous that, in time, John Outcalt found it necessary to travel back East to Gary, Indiana, to buy carriages for it. It was here in the Hoosier state that he met Miss Flora Johnston and talked her into visiting him out in the "wild West." She and her chaperone brother visited Gunnison the following summer on a sightseeing trip. John was thoroughly smitten by the pretty, dark-haired Miss Flora and, after her return to Indiana, they corresponded regularly. The romance apparently strengthened through the mail, and the two were married on Christmas day 1888 at the home of Flora's mother in Indiana. John brought his new bride back to Colorado to the homestead north of Gunnison, where he had already built a log dwelling, corrals, and barns. Flora was welcomed to a wide pretty valley with magnificent views of mountain ranges to the north; in summer, its large, flat floor grew thick with mountain hay, and the nearby sagebrush hills were flush with wildflowers.

Although cattle ranching was a gamble for the newly married Outcalts, it was a venture that allowed them to be somewhat anchored by the security of ownership of land. This ownership was granted to them by the government in return for the satisfaction of the requirements of the various early homestead acts. The beautiful, unmarked, undivided, unfenced Gunnison Country that the Richardson Colony had first found had begun to change its character now as acts created on paper on a desk in Washington, D.C., surfaced as genuine squares and rectangles on real land. John and Flora, like many other homesteading families, began adding on to the original homestead and building it up. The addition of more land meant more brush to be cleared, more ditches to be dug, more cattle to be bought, and endless work to be done. And along with the added work came the arrival of a family—four pretty little girls—Ramona, with the dark hair and eyes of

John and Flora Outcalt with the children, left to right: Vevarelle on her father's lap, Fernzelle, Delsie, and Ramona. Circa 1905.

Photo courtesy of the Soverin family.

her mother; then Fernzelle and Delsie, who were blond and resembled more their father; and finally, Vevarelle, who was born in September of 1902. She was a plump, blue-eyed, happy baby—pampered by her older sisters—her father's pride and joy.

Ranchers hope for sons who will grow up and learn to help with the heavy ranch work, but John Outcalt, having none, did the best he could by putting his four daughters to work—all of them having to help in order for the ranch to survive. The four girls did all they could to assist with calving and haying, performing all but the heaviest work and loving it. When they were older, they rounded up cattle in the fall, going out from the ranch to the hills to ride for days, bringing cows and calves home—even if it meant riding in a blizzard. They assisted with the feeding of the cattle in the winter, using the huge two-horse draft teams to break through the crusted snow to the hay stacks. The horses would then pull the huge loads of hay on a sleigh to the hungry cattle. Horses were an integral part of the operation of the Outcalt's ranch, and by the age of ten, Vevarelle was put to work training them.

The Outcalt sisters in Indian garb. Left to right: Fernzelle, Vevarelle, Ramona, and Delsie. Date unknown.

Photo courtesy of the Esty family.

Chapter Ten Vevarelle Esty of Gunnison

The Outcalt family and crew eating lunch in the hayfield in 1910.
Photo courtesy of the Soverin family.

The yearly harvest of hay—"haying"—was a family affair for the Outcalts and required the use of horse-powered machinery and the help of many hired men. Transient labor was in good supply during haying season, coming in on the railroad. A bunkhouse to sleep in and sparse wages were provided for the Outcalt's haying crew, along with great quantities of food—beef, potatoes, gravy, vegetables, pies, and cakes. Flora, with the help of whoever wasn't working in the hayfield, had one of the hardest jobs— cooking for the crew— from dawn to dusk, seven days a week, rain or shine. Vevarelle claimed that, at an early age, she proved her worth at jobs in the hayfield in order to escape from the endless tasks of the kitchen.

The Outcalt's early-day haying operation involved a tremendous output of horse and manpower. Their hay was first cut down by a mowing machine, or several mowing machines that were drawn by teams of horses. These machines had sickles that were lowered to the ground and cut five-foot swaths. Power came from the rotation of the mower's wheels, which turned gears and pitmans, or connecting rods, to move the sickle blades back and forth. Next a sulky rake, also pulled by a team, put the hay in windrows and bunched it to be picked up by a buck, bull, or sweep rake (all the same) and pushed to the stacker. The buck rake had large wooden pick-like teeth in front. Two men would receive the hay on the stack after it was dumped and distribute it evenly, making a bread-loaf shaped stack. Vevarelle and her sisters learned the skills of the hayfield early and put in long and tiring days along with the rest of the crew; in spite of the backbreaking work, they seemed to enjoy the process.

Ranch chores were important, but school was equally so. John Outcalt believed in education and was largely responsible for building the school that the girls attended nearby, which was called the "Paragon." This well-built structure survived for many decades and even now stands sturdy and tall with its round tower on the grounds of Gunnison's Pioneer Museum. Vevarelle had plans to continue with her education after graduation from high school, enrolling at Gunnison's college—the "Normal School"—now known as Western State College. But the death of John Outcalt immediately changed her plans for a college education.

The painful loss of Vevarelle's father, with whom she had a very close bond, left a void in her life that she felt could best be filled by taking care of the family ranch. Here she felt she was needed and she decided to stay on—a stay that lasted eighty years. The oldest of Vevarelle's sisters, Ramona, died around the same time as John Outcalt, a victim of the flu epidemic of 1918-19. Vevarelle later named a daughter after her. When Vevarelle's mother, Flora, later died, the ranch was divided among the three surviving sisters. Delsie married and moved away, Fernzelle married and settled on another location on the Outcalt homestead, and Vevarelle stayed on at the home place in her parent's log house. When she married Eugene Esty, a local rancher and miner, the two set up housekeeping at the family homestead and continued to run the ranch while raising a family of three children. Through the arrivals of her children, Vevarelle carried on as usual with the managing and chores of the ranch, which by now she was so well accustomed to that it was second nature. She nursed babies, changed diapers, pitched hay, and branded calves—all in a day's work.

The death of Vevarelle's husband in 1946 and, later, the death of a son necessitated her continued involvement with the running of the ranch for most of her life, totally on her own and doing much of the work herself. She branched out into other ventures from time to time when there were lean years. During the Depression, she all but supported the family with a truck garden. She and her son would drive trucks from Gunnison to Grand

Junction, stopping at restaurants along the way to make deliveries. Vevarelle seemed to have a special gift for gardening. After the Depression, she kept the huge garden going, and, for more than sixty years, she raised her luxurious rows of onions, carrots, squash, potatoes, beans, cauliflower, broccoli, beets, turnips, rutabagas, mustard greens, and dill weed. She canned or pickled the produce, too, in her spare time from ranch duties.

At the time Eugene died, Vevarelle was managing around 2,500 acres, gardening, and raising three young children. She was mother, father, ranch manager, cook, and cleaning lady. The difficulty was staggering. To add to it, three years after becoming a widow, her home, built by her father in the 1880s, burned to the ground. Vevarelle was barely making ends meet and didn't have the money to have blueprints drawn up for a new home. In typical Vevarelle style, she drew them up herself and then proceeded to do much of the construction of the new home, gambling that winter wouldn't come before she and the children finished the roof. All assisted in shoveling out several inches of snow that had fallen on the newly finished floor when winter arrived ahead of time. But it worked out somehow, and Vevarelle lived in the house she had built for the rest of her life.

Vevarelle never wavered through all these difficult years of managing a ranch and parenting alone. In fact, her work ethic, born of her pioneer beginnings, was carried out in as many civic duties as she could manage, as well. She was chairperson of the Gunnison County Farm Bureau for fifty-three years; received numerous awards from her fellow ranchers—all men—from the Gunnison County Stockgrowers Association; was presented the "Rancher of the Year" award from Gunnison's Chamber of Commerce in 1971; was named once as Conservationist of the Year by the local Soil Conservation District; and . . . was cited for service by the Colorado Cattlemen's Association. All these awards probably paled in comparison to the one thing she was proudest of. She was, in her later years, when most people are retired, getting more tonnage of hay and producing larger cattle than the Gunnison County average. Excellence is often its own reward, and Vevarelle's particular hard-won brand of excellence on the grounds of her own ranch may have meant more to her than any public recognition.

All through her long life, Vevarelle's successes had been hard won. From the time she was a young girl, she had a habit of taking on tasks that were considered difficult, unconventional, or downright impossible for females.

According to one of Gunnison's old-time cattlemen, she competed in the town's early-day rodeos by riding broncs—hardly an acceptable activity for a young lady of that time. Other girls her age were riding in the Cattlemen's Days parade in wagons festooned with flowers and wearing lacy dresses and pretty bonnets. Not Vevarelle—she insisted on riding down a bucking horse. Later in life, when she had sole responsibility for the ranch, she was up at dawn and ready to take on anything it required. She built and repaired her own barbed-wire fences, a backbreaking job for a man, and even more of a challenge for a woman. Not content with her hay meadow's natural state, as some local ranchers were, Vevarelle disked up a portion of hers and planted a combination of oats and hay and realized a good crop from it. Later, she somehow established connections to Japan to sell her high-quality hay. Every blade of grass out on her range was utilized by her cattle and every stalk of hay was mowed down in her hayfields, including that which was customarily left on ditch banks. Vevarelle was always on the move, either on a horse or in her red Jeep. When she wasn't ranching, she was gardening or attending agricultural meetings. Always a rider of big horses,

The surviving Outcalt sisters in later years in a Cattlemen's Days parade. Left to right: Vevarelle, Fernzell, and Delsie.

Photo courtesy of the Esty family.

she solved the difficulty of getting on one when she was older by placing a tree stump by each gate she had to open, giving herself a little assistance in getting up. Work was her life—her occupation from sunup to sundown, seven days a week.

Vevarelle had only sporadic hired help, and those men who worked for her or leased some of her land came to have the utmost respect for her. It is reported, however, that in the early days, a few of her male counterparts waited patiently for her to fail, thinking they could pick up a tremendous ranch pretty cheaply when it happened. They waited a long time, and then observed as she squelched their ambitions by pulling the ranch she had inherited out of bankruptcy. She just simply kept on working, kept on amazing people, and never looked back. An admiring male contemporary of hers stated that when Vevarelle was eighty years old, he offered to rope a bull in a corral for her. Her reply? "Just because I'm eighty years old doesn't mean I can't rope a bull." Then she proceeded to do just that.

Vevarelle retired in 1993, feeling that, in her nineties, she needed to slow down. She sold her ranch at this time but continued to live in her home and act as an advisor to the new owners and managers. She remained intensely proud of the ranch she had taken care of for so many years. With great delight, she would point out an ancient barn that her father had built and built well—three stories high—the first floor for livestock, the second for storing hay, the third for storing grain, and, on the very top, boxes for pigeons.

Vevarelle had obvious ailments at an advanced age, but she offered no complaint about them. She often expressed curiosity as to why some people live so long compared to others. She seldom mentioned her children and rarely mentioned her husband, long dead. Her thoughts went back again and again to the early days to "Daddy," a man for whom she never lost her great affection. She recalled as if it was yesterday, his arrival home to four young girls who ran to him when he arrived at the ranch gate with a team of horses, and lifted them up one by one to the horses' backs to accompany their father along home.

Like any other old rancher, by the time Vevarelle reached the end of her life, character was etched on her leathery face. The strength in her face and in her bearing, and the intelligence and kindness in her eyes, spoke of an unburnished kind of beauty. She died at the age of ninety-seven, having been the strong force that carried on the pioneer Esty Ranch to well beyond a century and having been a living example of her belief that "you don't find true happiness in life until you find it in your work." Vevarelle passed on the pioneer blood of John and Flora Outcalt to three children, seven grandchildren, thirteen great-grandchildren, and two great-great-grandchildren.

Vevarelle on one of her tall horses in 1985.
Photo courtesy of the Esty family.

Bibliography

CHAPTER ONE
BOOKS

Brown, Dee. *The Gentle Tamers—Women of the Old Wild West*. Lincoln and London: University of Nebraska Press, 1958.

Brown, Margaret Duncan. *The Shepherdess of Elk River Valley*. Denver: Golden Bell Press, 1967.

Haley, J. Ebetts. *Charles Goodnight—Cowman and Plainsman*. Norman: University of Oklahoma Press, 1949.

Jester, Norton. *A Woman's Place—Yesterday's Woman in Rural America*. Golden, Colorado: Fulcrum Publishing, 1996.

Meyers, Sandra L. *Westering Women and the Frontier Experience*. Albuquerque: University of New Mexico Press, 1982.

CHAPTER TWO
BOOKS

Decker, Peter R. *Old Fences, New Neighbors*. Tucson: University of Arizona Press, 1998.

Gregory, Doris. *The Town That Refused to Die—Ridgway, Colorado 1890-1991*. Second Edition. Ouray, Colorado: Cascade Publications, 1992.

Stanton, Henry. *Trail Dust*. Grand Junction, Colorado: USA Precision Printing, 1990.

Steinel, Alvin T. *History of Agriculture in Colorado from 1858-1926*. Ft. Collins, Colorado: The State Agricultural College, 1926.

NEWSPAPERS

The Daily Sentinel. March 8, 1981.
The Denver Post. May 29, 1983.
The Farm Quarterly. Fall, 1960.
The Montrose Daily Press. November 7, 1979, March 21, 2003.
The Telluride Times. November 8, 1979.

INTERVIEWS

Ken Reyher and:
> Wesley Henshaw, January 8, 2004.
> Dick Swyhart, November 12, 2003; December, 2003.

Judy Buffington Sammons and:
> Nick Gray, February 19, 2004.
> Delbert Frasier, August 22, 2004.
> Willma Potter, January 16, 2004.
> David and Mary Wood, December 30, 2003; July, 2004.
> Mario Zadra, March 7, 2004; July, 2004; August, 2004; September, 2004.

CHAPTER THREE
BOOKS

Marr, Josephine Lowell and Joan Marr Keiser. *Douglas County—A Historical Journey.* Gunnison, Colorado: B&B Printers, Inc., 1983.

Tanner, Ogden. *The Ranchers.* Alexandria, Virginia: Time-Life Books, 1977.

PERIODICALS

Frickel, Marijean. "Tweet Kimball: Citizen of Douglas County—Citizen of the World." *Colorado Mature Lifestyles,* April-May, 1999.

Mullan, Joyce. "A Fortress for a Queen: Tweet Kimball's Sedalia Castle." *Colorado Homes and Lifestyles,* May-June, 1981.

OTHER RESOURCES

Schler, David. Video Recording: *"A Jewel in the Rockies—The Story of Cherokee Ranch,"* Douglas County Television, 1999.

NEWSPAPERS

Castle Rock Chronicle. December 10, 1996.

The Denver Post. January 16, 1999.

Douglas County News Press. December 14, 1983; May 8, 1996; January 20, 1999.

Rocky Mountain News—Sunday Magazine. June 1994.

The Weekly News & Castle Rock Chronicle. January 20, 1999.

INTERVIEWS

Judy Buffington Sammons and:
- Charles Kirk Jr., April 15, 2005, Castle Rock, Colorado, by telephone.
- Marcia Wright Landwehr, April 16, 2005.
- Donna Wilson, Executive Officer, Cherokee Ranch and Castle Foundation, October 8, 2005, by telephone.

CHAPTER FOUR
BOOKS

Bailey, William McCrea. *Fort Uncompahgre.* Silverton: *The Silverton Standard and The Miner,* 1990.

Brown, Margaret Duncan. *The Shepherdess of Elk River Valley.* Denver: Goldon Bell Press, 1967.

Doherty, Deborah. *Delta, Colorado—The First 100 Years—A Centennial History,* First Edition. Delta, Colorado: *Delta County Independent,* 1981.

Marshall, Muriel. *The Awsome 'Dobie Badlands.* Montrose, Colorado: Western Reflections Publishing Company, 2000.

Marshall, Muriel. *Island in the Sky.* Montrose, Colorado: Western Reflections Publishing Company, 1999.

Marshall, Muriel. *Red Hole in Time.* College Station: Texas A&M University Press, 1988.

Reyher, Ken. *High Country Cowboys—A History of Ranching in Western Colorado.* Montrose, Colorado: Western Reflections Publishing Company, 2002.

NEWSPAPERS

Delta County Independent. January 8, 1937.

OTHER RESOURCES

District Court—County of Delta. Petition for Probate of Will of Genevieve Hardig. April 1, 1974.

"Hotchkiss and Crawford 1881-1910." Brochure of Hotchkiss Public Library. 1976.

PERSONAL INTERVIEWS

Judy Buffington Sammons and:
    Bob Barns, November 24, 2004.
    Larry Davidson, November 1, 2004.
    Gordon Hodgin, October 25, 2004.
    Gary Robbins, November 30, 2004.
    Ben Walker, November 4, 2004.

CHAPTER FIVE
BOOKS

Bennett, Edwin L., and Agnes Wright Spring. *Boom Town Boy.* Chicago: Sage Books, 1966.

Halka, Chronic, and Felicie Williams. *Roadside Geology of Colorado,* Second Edition. Missoula, Montana: Mountain Press Publishing, 2002.

Huston, Richard C. *A Silver Camp Called Creede—A Century of Mining.* Montrose, Colorado: Western Reflections Publishing Company, 2004.

La Font, John. *Fifty-eight Years Around Creede.* New York: Vantage Press, Inc., 1971.

Wright, Mabel Steele. *River Ripples.* Gunnison, Colorado: B&B Printers, 1959.

NEWSPAPERS

The Lake City Silver World. July 30, 1993.

The San Luis Valley Historian, Volume XXIV, Creede Centennial Issue. November 1, 1992.

Silver Thread—Colorado Scenic and Historic Byway, A Summer Guide. Publication of *The Lake City Silver World.* Summer, 2000.

INTERVIEWS

Judy Buffington Sammons and:
    Jerry and Polly Bates, February 12, 2005.
    Grant Houston, (loan of personal files on the Steele family, July 30, 1993.)
    Bill and Sue Knowles, February 5, 2005.
    Margaret Lamb, March 8, 2005, Creede, Colorado, by telephone.
    Dorothy Steele, February 27, 2005.
    Ed Wince, February 18, 2005.

CHAPTER SIX
BOOKS
Propst, Nell Brown. *Forgotten People—A History of the South Platte Trail.* Boulder, Colorado: Pruett Publishing Company, 1979.

Tanner, Ogden. *The Ranchers.* Alexandria, Virginia: Time-Life Books, 1977.

Walters, Hildred and Lorraine Young. *More Prairie Tales.* Printed by Business Farmer, Scottsbluff, Nebraska, 1976.

PERIODICALS
Denhardt, Bob. "Marshall Peavy." *Western Horseman Magazine,* March 1990.

NEWSPAPERS
Appaloosa News. August, 1980.
Greeley Tribune. April 21, 1981; May 22, 1984.
Maverick Press. March, 2004.
Sterling Journal—Advocate. March 31, 1975; August 19, 1987; January 29, 2000.

INTERVIEWS
Judy Buffington Sammons and:
- Jennie Mai Bonham, November 12, 2005, Cheyenne, Wyoming, by telephone.
- Marsha Witte, October 6, 2005, Peyton, Colorado, by telephone.

CHAPTER SEVEN
BOOKS
Tanner, Ogden. *The Ranchers.* Alexandria, Virginia: Time-Life Books, 1977.

PERIODICALS
Sykes, Velma West. "Pioneer Life in the San Luis Valley— As told by William Hanson." *Colorado Magazine,* July, 1940.

OTHER RESOURCES
Hansen, Julie. "The Peter Hansen Family Story." Personal files of Eileen Hansen Burt. July, 2002.

"Monte Vista Downtown Historic District." A pamphlet published by the Monte Vista Historical Society, 110 Jefferson Street, Monte Vista, Colorado.

NEWSPAPERS

Alamosa Courier. March 3, 1917.

Alamosa Empire. October 17, 1923.

Colorado Prospector—Colorado History From Early Day Newspapers. May 1986.

Courier Leader. June 16, 1923.

Daily Courier. March 4, 1951.

Pueblo Chieftain. September 2, 1983.

San Luis Valley News. June 9, 1923.

Valley Courier. November 19, 1956.

INTERVIEWS

Judy Buffington Sammons and:
   Harry Becker, Jr., June 22, 2005.
   Eileen Hansen Burt, May 2, 2005.

CHAPTER EIGHT
BOOKS

Brown, Larry K. *Coyotes and Canaries—Characters That Made the West Wild . . . and Wonderful.* Glendo, Wyoming: High Plains Press, 2002.

Burroughs, John Rolfe. *Where The Old West Stayed Young.* New York: William Morrow and Company, 1962.

McClure, Grace. *The Bassett Women.* Athens, Ohio: A Sage Book of Swallow Press—Ohio University Press, 1985.

Monnett, John H. and Michael McCarthy. *Colorado Profiles—The Men and Women who Shaped the Centennial State.* Evergreen, Colorado: Cordillera Press Inc., 1987.

Vandenbusche, Duane, and Duane A. Smith. *A Land Alone.* Boulder, Colorado: Pruett Publishing Company, 1981.

PERIODICALS

Bassett, Anne Willis. "Queen Anne of Brown's Park." *The Colorado Magazine*, April, 1952.

CHAPTER NINE
BOOKS
Halka, Chronic, and Felicie Williams. *Roadside Geology of Colroado,* Second Edition. Missoula, Montana: Mountain Press Publishing, 2002.

Marshall, Muriel. *Island in the Sky.* Montrose, Colorado: Western Reflections Publishing Company, 1999.

Sammons, Judy Buffington. *Tall Grass and Good Cattle—A Century of Ranching in the Gunnison Country.* Gunnison, Colorado: Western State College Foundation Inc., 2003.

OTHER RESOURCES
Gunnison County Stockgrowers 1894 -1994. 100 Year Anniversary Issue. B&B Printers. Gunnison, Colorado, 1994.

Rangelands Magazine—Twenty-five Years of Information. Society for Range Management Publication, 2003.

INTERVIEWS
Judy Buffington Sammons and Barbara East, July 24, August 20, 2005.

CHAPTER TEN
BOOKS
Sammons, Judy Buffington. *Tall Grass and Good Cattle—A Century of Ranching in the Gunnison Country.* Gunnison: Western State College Foundation, Inc., 2003.

OTHER RESOURCES
Personal Files of Vevarelle Esty. Gunnison, CO. 1995.

NEWSPAPERS
Crested Butte Chronicle and Pilot. November 14, 1986.
Gunnison Country Times. October 1, 1981; September 30, 1999.

INTERVIEWS
Judy Buffington Sammons and:
    Jerry Bates, February 12, 2005.
    Vevarelle Esty, August, 1993; October, 1994.
    Suzanne Esty, October, 2004.

About the Author

Judy Buffington Sammons grew up on a ranch in the Ohio Creek Valley of Gunnison, Colorado, where her father and grandfather raised registered Hereford cattle. She and her family still maintain ownership of part of the original Buffington Ranch, working and enjoying it during the summer months. Judy holds a Master of Arts Degree in Education and has recently retired from a thirty-year teaching career in the field of Adult Education. Her first book, *Tall Grass and Good Cattle: A Century of Ranching in the Gunnison Country*, was commissioned and published by the Western State College Foundation, Inc. Judy resides in Gunnison, Colorado, is married, and the grandmother of two.

If you enjoyed *Riding, Roping, and Roses*, you may like reading these other books from Western Reflections Publishing Co.:

Maggie's Way: The Story of a Defiant Pioneer Woman by Lucinda Stein

Bess: A Woman's Life in the Early 1900s by Carol McManus

Ida: Her Labor of Love by Carol McManus

High Country Cowboys: A History of Ranching in Western Colorado by Ken Reyher

Silverheels by Tara Meixsell

Colorado Mountain Women: Tales From the Mining Camps by Sherie Schmauder